AF349012

Beyond Simplicity

Why Studying Complex Concepts is Essential

Daniel Wieser

Also by Daniel Wieser

Speed Reading Genius

Daniel Wieser

Beyond Simplicity

Why Studying Complex Concepts is Essential

SLV

• Wien •

Dedicated in honor to mathematicians and engineers alike.

To Ulli

Table of Contents

Preface

"Beginnings are always difficult
in all sciences."[1] – Karl Marx

After multiple readings of and corrections of this book, I found myself rewriting it once again. Surprisingly, in the end, everything made even less sense than it initially appeared. There could be several reasons for this, which I'll briefly shed light on. Firstly, the subject matter itself is complex, focusing on the complexity of various areas of knowledge and the purpose of learning them. Additionally, I aimed to create an extensive work that drew connections to numerous other books through direct quotations. At the time of writing, these quotes provided me with inspiring insights. However, I later realized that without immediately clarifying the specific ideas that sparked during those moments, these quotes only added to the complexity and confusion. Moreover, I attempted to present my work in a manner that would be understandable to the reader, but unfortunately, I failed miserably. I tried to impose an organizational system that wasn't originally designed in this way. Now, feeling lost in understanding how everything fits together harmoniously, I decided to rewrite it once again. I acknowledge that the lack of understanding in conveying my intended message is a criticism I frequently

---

[1] Preface to the First Edition 1867, republished in Marx (1990)

encounter, and I anticipate hearing it again. Nevertheless, I have come to accept it, as there will be a few who will comprehend my words. It is for those individuals that I have written this book. However, I have also written it for those who wish to embrace this mode of thinking, those eager to acquire new knowledge and perhaps willing to make the effort to read this book. If you find yourself among these individuals, I express my gratitude for your dedication. Rest assured, it will undoubtedly be rewarding. The following work is a compilation of my earlier theories on the *Universalgenie* as well as later ones which focus on the complexity of the matter that the renaissance man or woman has to deal with. All of them are contributions to the retroactively construction of a theory which hopefully will be evident after reading this book completely. This book explores the usefulness of acquiring knowledge in complex subject matters, supported by both empirical studies and personal experiences. It also incorporates an autobiographical perspective, recounting my own background and insights. Although scientific sources were not always available, I made a substantial effort to integrate relevant research whenever possible. Where this has not been possible, I ask the reader for their indulgence. In some respects, I did not have the patience necessary to wait until there are enough suitable studies officially available that adequately reflect my own state of knowledge, which is why I published my own experience without further ado, as I do not want to withhold my insights from the reader any further. This book was born out of the motivation to stimulate interest in complex topics and to improve the initiative for people to engage more with complex ideas and learn to understand them. This is not always easy and requires a lot of time and effort – an effort that many may not be willing to undertake, or may not perceive as necessary. However, it is precisely this effort that is essential, and it is this very idea that this book aims to convey. In the foundation of this particular discourse, the general aim is to provide the reader with a method so that they have the necessary means to comprehend the most complex concepts and themes inherent in this world. This treatise is concerned not only with the *complexity* of concepts but also with their *comprehension*. It is based on a logic that aims to represent the reality in its integral and complex form rather than dissolving it. The purpose of this work is to reveal the meaningfulness of learning complex entities and abstract concepts through their cognition by showing the importance of delving deeper into various subject matter and going beyond simplicity. It cannot be made recognizable by the limitations of simplification but requires exposure to a variety of different complex *concepts*, *objects*, *systems*, or *phenomena* that possess multiple interrelated components or layers. The richness of the concrete substance can only be perceived through its immanent content, which otherwise remains incomprehensible. This act of learning, although it often simplifies the matter, ultimately reveals the origin – the complex entity itself – rather than providing a comprehensive truth about the subject matter. The scientific character of this work should not be discussed here but rather revealed to the reader through their engagement with this discourse on the complexity of concepts. Necessary supplements are provided as footnotes, including references to various sources. The writing style may change from complex-analytical to simple and understandable at times. This intentional shift contributes to the unique character of the book, driven by the author's tendency to follow, reconstruct, and adapt their own trains of thought. One might think it is no easy task, but in many respects, this approach has been successful. The inherent content of this work and its theoretical approach are based on dialectical materialism as a philosophy of knowledge. Through theoretical modifications and analogous considerations, we employ mental representations to create assumptions that accurately reflect the content of this work as a whole. The form and essence of this composition significantly complicate the accessibility of the subject matter. However, it is through this complexity that the true meaning of the work is revealed. A thorough engagement with the complexities is essential to grasp the essence of this work and others like it. Genius does not create with the intention of conveying a specific truth or creating an object of beauty. Instead, what they create is a genuine expression of themselves, a reflection of their own reality and how they perceive the world. This work, in the truest sense, is such a reflection. When I write, I often have people in mind who I write this book for. *Beyond Simplicity* is a book that was written for someone who does not exist anymore – my past self. Nevertheless, I hope that some other people might read it in the future and gain the motivation and understanding of the importance of dealing with complex subjects. One of them being my dear Ulli, who has shown increasing interest in my writing and has taken the time to carefully read the first draft. I was happy to hear that this then motivated her to pursue a Master's degree at the university of applied Sciences in St. Pölten, showing me that this book truly has the power to transform lives.

Daniel Wieser,
June 2023

Acknowledgements

First and foremost, I would like to express my heartfelt gratitude to my grandfather, whose unwavering passion for engineering served as a profound inspiration for me. Although I deeply regret not having had the opportunity to discuss my academic pursuits in engineering with him, his influence remains an indelible force in my life.

Next, I would like to extend my sincere appreciation to my parents, Ulrike and Franz Wieser, for their constant encouragement and unwavering support throughout my journey in the field of engineering. They have never hesitated to delve into any book I have written and presented to them, regardless of its quality, displaying a genuine interest in my intellectual growth.

Due to unforeseen circumstances, the brilliance of my former schoolmate Paul Wedrich, whose exceptional achievements in mathematics surpass my own aspirations, has been increasingly motivating for me to study pure mathematics and pursue my intellectual development further in this field[2]. His remarkable talents and accomplishments serve as a constant reminder for me to further explore the vast realm of possibilities within mathematics.

I am immensely grateful to my dear friend and fellow student, Arthur Klammer, whose invaluable assistance has

---

[2] Mathematical pun intended.

been instrumental in navigating the intricacies of my business informatics classes.

I also owe a debt of gratitude to my friend Christoph Lenz, with whom I have engaged in countless insightful conversations on various mathematical topics. This not only gave me a better understanding of the topic but also the courage to deal with complex topics in more detail.

I extend my thanks to my diverse circle of friends, each of whom has pursued remarkable fields of study. Their unique academic endeavors have continually inspired and motivated me to push the boundaries of my own intellectual pursuits.

"Il me vient quelques fois tant de pensées le matin dans
une heure pendant que je suis encor au lit, que j'ay besoin
d'employer toute la matinée et par fois toute la journée et
au de là, pour les mettre distinctement par écrit."[3]

"Sometimes in the morning, while I am still in bed, I have
so many thoughts that come to me in such abundance
that I need to spend the entire morning, and sometimes
even the entire day and beyond, to put them down in
writing distinctly."

– Gottfried Wilhelm Leibniz

[3] Bodemanm, 1889

Introduction

Perhaps you have already asked yourself this question: Why should I engage with complex subjects? I attempt to provide a possible answer to this question with this book. It is a collection of insights that I have gained over time through experiences with various topics. If someone had asked me personally ten years ago if pursuing a scientific degree was an option, I probably would have responded that it may not lead to substantial financial gain.

However, that was a very short-term perspective on life, as will be shown, and I can't blame anyone but myself. Perhaps I wouldn't have even understood if an enthusiastic scientist had told me about their work. Enthusiasm cannot be grasped; one must experience it firsthand. Now that I have had the privilege of experiencing this enthusiasm, I want to at least try to share it with you. I want to motivate you to engage with more complex subjects, even if the immediate benefits of such an endeavor may not be evident.

Part 1

Motivation & Benefits

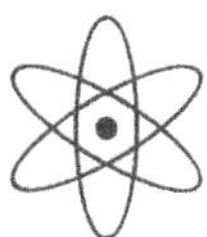

Why Studying Complex Concepts is Important

"Learning is the only thing the mind never exhausts,
never fears, and never regrets."
Leonardo Da Vinci

The Problem Today

"The science of to-day can give us no certain answer."[4]

Let's first highlight the problem that is currently inherent in our society, that is in terms of dealing with complex concepts in general. The symbol featured on the cover is a stylized depiction of the atomic model proposed by Niels Bohr in 1913. Widely recognized and frequently used in popular media, it is emblematic of the complex scientific concepts that capture the public imagination. However, despite its iconic status, this model is now considered outdated in light of the more sophisticated concept of atomic orbitals. In many ways, the symbol serves as a fitting introduction to the theme of this book:

The hazards of
oversimplification in science.

AND

The benefits of studying and
understanding complex concepts.

[4] Borgstroem, 1921

In order to make scientific concepts more accessible to the general public, they are often simplified which takes away from the beauty of the complexity of it as well. Sometimes these concepts are oversimplified to the point where they no longer accurately reflect the current state of knowledge. This approach may generate greater public interest in science, but it also risks oversimplifying the subject matter in a way does no good to science itself nor the person who wants to further engage with it. It may also be encouraging individuals to feel that they understand scientific concepts when, in fact, they do not – and people will ultimately fail to appreciate the full complexity of scientific concepts. And when complex concepts are oversimplified to the point that everyone thinks they understand them, where is the motivation to explore science more deeply? This is a pity because science has a lot to offer for those who take the time to engage with it. Science holds all wealth of knowledge and potential for advancement. If you are content with superficial simplification, this often results in a shallow understanding, a lack of depth – necessary for a comprehensive grasp of complex scientific concepts –, and a lack of grit in other areas of life. During the COVID pandemic for example, it has become apparent that many individuals have accepted theories without critical examination or further research, resulting in the spread of fake news.[5] While it is not necessarily problematic to read such information – if one has the ability to think independently critically, and thoroughly, it's easy to detect false information as such[6] –, it is crucial to approach such simplifications with a

[5] Gupta et al., 2022
[6] Escolà-Gascón et al., 2021

healthy degree of skepticism. Solely trusting the experts, without taking the time to engage with the material yourself, is not a viable option either:

> "The SARS-CoV-2 pandemic made it clear that disagreements among experts may extend even to such seemingly simple decisions as wearing a face mask."[7]

Terminology used in scientific fields can be particularly complex and difficult to understand at first glance, necessitating a concerted effort to engage with the material. However, people do not always take this effort but rather form an opinion right away – this can then lead to problems despite the lack of knowledge. Is this the future in which people want to exchange ideas and make decisions – and most of all, is this the future you want for yourself? Should we not strive to expand our knowledge in several scientific areas to know more in general? Of course, it is impossible to be an expert in all fields, but that is not the point here. Logical thinking skills are appropriate in any endeavor. Only after thoroughly engaging with the material can one make informed decisions, and a lack of understanding and information can lead to misguided judgments and actions. To address this issue, we will explore the many reasons why studying science and engineering is vital, why it is necessary to focus on learning complex concepts, and why having an eagerness to learn more is essential for success in these fields (*Why You Should Focus On Science Today*, page 40). From advancing technology and innovation to improving critical thinking and problem-solving skills, we will dive deep into

---

[7] Smil, 2022

the benefits of learning complex concepts and how it can help us achieve our goals and create a better future – not only for ourselves, but for the world at large.

> "The complexity of material has likely increased dramatically, yet our reading skill has remained the same."[8]

So, if you want to stay ahead of the curve and make a meaningful impact in your field, continue reading on the next pages and join me on this journey of discovery to learn why learning complex concepts is the key to unlocking your full potential.

> "[…] it is difficult to accept that the best learning road is slow, and that doing poorly now is essential for better performance later."[9]

Before we get started with the book, however, I would like to share a quote along the way from an author I admire highly. As you will see throughout the book, I have quoted many passages – such as the one following – entirely, because I find that parts of it, taken out of context, would not have the same effect to the reader and it might not be possible to grasp exactly what I wanted to say:

> "I am preaching the message that, with apparently only one life to live on this earth, you ought to try to make significant contributions to humanity rather than just get along through life comfortably that the life of trying to achieve excellence in some area is in itself a worthy goal for your life. It has often been observed the true gain is in the struggle and not in the achievement – a life without a struggle on your part to make yourself excellent is hardly a life worth living. This, it

[8] Kwik, 2020
[9] Epstein, 2019

must be observed, is an opinion and not a fact, but it is based on observing many people's lives and speculating on their total happiness rather than the moment-to-moment pleasures they enjoyed. Again, this opinion of their happiness must be my own interpretation as no one can know another's life. Many reports by people who have written about the "good life" agree with the above opinion. Notice I leave it to you to pick your goals of excellence, but claim only a life without such a goal is not really living but merely existing_ in my opinion. In ancient Greece, Socrates (469-399 BC) said: The unexamined life is not worth living."[10]

This is truly a call to engage with complex topics - such as science, engineering, mathematics, and so on. Yet, many people would rather watch a simple YouTube video. Perhaps the problem of oversimplification is that fewer people today would engage with such topics or perhaps there are fewer opportunities to do so. It seems the opposite is true, as we will see on the following pages.

> "Today, there are more scientists, more funding for science, and more scientific papers published than ever before."[11]

The interest in science is ever increasing and the growing interest in science has led to an increase in the number of students pursuing science-related fields. Due to science's growing significance in forming our understanding of the world and solving difficult challenges, the study of science has become more and more relevant for people, companies, and governments all around the world especially in the addressing of today's complex problems – problems that the average person would typically not encounter in daily life. Therefore,

[10] Hamming, 2020
[11] Collison and Nielsen, 2018

the ordinary citizen will not deal with such problems unless they see the need in the big picture. How true, the ordinary citizen will not deal with something out of the ordinary – either because they never have to deal with it due to challenges appearing in their own life, or because they will never deal with it due to their own interest. What a pity, because despite the fact that science is more present today, skepticism in science is increasing in uneducated individuals – or at least, that is what one would assume:

> "the assumption that science skeptics are ignorant or stupid is not supported by the evidence which seems to indicate, for example, that highly educated people are no less likely to doubt science than people with lower levels of education."[12]

It does not take a genius to understand that many major problems are solved and can be solved with the help of science.[13] But thankfully, individuals are becoming more aware and curious about science as they recognize its relevance to their everyday lives and its potential for improving human well-being overall.[14]

More and more organizations are seeking to incorporate scientific findings into their practices and strategies, recognizing the value of data-based decision-making.[15] And governments, on the other hand, are investing heavily in scientific research and education[16] to promote innovation, economic growth, and national security. Additionally, the

[12] Contessa, 2022
[13] Ho & Sculli, 1997
[14] Wang et al, 2021
[15] Miragliotta et al, 2018
[16] Maes et al., 2011

increase of global challenges, such as the recent pandemic, have further driven the need for scientific research and development.[17] The demand for scientific solutions to these complex issues has contributed to the increasing interest in science on a global scales. The accessibility of scientific information has also enabled individuals to become more informed[18] and aware of the potential benefits and implications of scientific advancements. Individuals are now able to access scientific studies, articles, and research papers from anywhere in the world. Of course, social media also contributes to this. The further growth of social media and online communication platforms has made scientific information more accessible to a broader audience. And yet, more information does not necessarily mean more knowledge. How do we acquire more knowledge? To learn more, we have to take the effort to deal with difficult topics (learn more about *Intellectual Challenges* on *page* 60).

> "[...] innovative learning is a necessary means of preparing individuals and societies to act in concert in new situations, especially those that have been, and continue to be, created by humanity itself. Innovative learning, we shall argue, is an indispensable prerequisite to resolving any of the global issues."[19]

There are always people who take the difficult path: They deal with the complex contents and take the time and effort to learn these things, regardless of whether they will earn a lot of money with it later on or not. Why is it that some people take this effort? Is it scientific curiosity or the curiosity in

[17] Yun, 2023
[18] Purcell & Rainie, 2014
[19] Botkin, 1979

dealing with difficult topics? Why should one bother to study mathematics and engineering? Shouldn't one perhaps rather study something simple that one knows one can complete?

> "In recent years, Science and Technology student numbers have been increasing in absolute terms, but decreasing in relative terms."[20]

You might think that there aren't enough ways for people to get interested in science, but that is not entirely correct. There has, in fact, been a notable increase in the availability of informal and formal science resources in recent years. What has been particularly relevant for the interest in science, is the informal science education, such as museums and popular non-fiction literature:

> "Informal science education was first provided for adults in the nineteenth century [...] The development of the notion of 'museum' and the availability of cheap paper-based publications were major factors in extension of the genre, first into family-based education, and then directly into the education of school-age students."[21]

Not only that, but with the emergence of new technological platforms, there has been a significant expansion of resources beyond traditional media. The internet, in particular, has been a game-changer, providing access to scientific papers and other materials. Furthermore, video-sharing websites such as YouTube have become a valuable source of science tutorials. Digital tools are helping students of all ages:

[20] OECD, 2006
[21] Gilbert, 2015

"Digital tools help students listen to math classes and respond by typing, scribbling, or dictating their remarks. As the school year progresses and the years pass, digital technology will be employed to make math education through practice more accessible to all students, regardless of ability or learning environment. These technologies also give pupils the option of studying or demonstrating their math skills in various ways. Because the learning environment is more dynamic than it has ever been, today's students are very different from those for whom the educational system was created. As technology improves, classrooms are being remodelled and reimagined in numerous ways to meet the increasing expectations of modern digital learners."[22]

In addition to these online resources, there are also various providers, such as EdX, that offer science courses and other educational materials for free. Platforms like Coursera also provide an opportunity to acquire certificates for a nominal fee. Udemy courses, meanwhile, offer relatively inexpensive options for those seeking to gain knowledge in science-related fields. The variety of options available today makes it easier than ever for people to develop an interest in science, but it is important to carefully evaluate the content available and ensure that it meets your own standards. But this has not always been the case...

"the lack of young people deciding to study technical and scientific fields becomes a crucial social problem. This dangerous trend may result in shortages of skilled researchers and technicians involved in the highly sophisticated research, which is at the beginning of all product-developing chains resulting in sales of high added value products."[23]

[22] Haleem et al., 2022
[23] Sládek et al., 2011

Even if this text quotes an article that was published ten years ago, there is substantial evidence to suggest that there exists a dearth of scientific education in several regions and amongst certain populations. In some countries, there may be a lack of adequate funding for scientific education – in others, there might be a lack of interest in science despite sufficient resources for educational institutions and more opportunities than ever for students. I do not want to go into detail how certain countries should enforce their education policies, nor is this an analysis of science education around the world. I am referring mainly to our latitudes and therefore to the lack of interest in that very country I am living right now. Because it seems that the

> "Disinterest in science in Austria is clearly more pronounced than systematic skepticism or lack of trust in science."[24]

One may be pleased, however, that it is only the disinterest, not the *trust*, that has decreased. Yet, this is not surprising and exactly the reason why I wanted to write this book you are reading right now. Nevertheless, a lack of interest in science and engineering is a crucial issue that needs to be addressed since it can adversely affect the society's interest in pursuing a higher education and a scientific career, which then further affects progress and development of that society. But what are those reasons why people are not *that* interested in science? I speak from my own experience here, and I know exactly why studying a science did not really interest me in the past. What was important to me at the time

---

[24] Standard, 2023 - *"Desinteresse an Wissenschaft in Österreich ist deutlich ausgeprägter als systematische Skepsis oder mangelndes Vertrauen in Wissenschaft."*

was earning myself a living – i.e. making money. This is because money is a simple concept that is easy to understand.

> "Humanly speaking, someone who thinks only of getting rich is living his life subjected to a thing, namely money."[25]

Once you get that, it is obvious to want to learn something that will help us to earn more money. We understand that we need to learn because we can comprehend what that goal – make more money – means. However, the meaning of such a goal has only manifested itself – that is, our conception of why such a goal is important – because we have understood, given the circumstances, what more money would be useful for. Someone who does not know the concept or meaning of money will not understand that and will not take the time and effort to pursue the earning of it. The acquisition of money is often used as an excuse to generate wealth as the highest possible achievement in one's life. If we have achieved that, then we should be able to focus on higher intellectual pursuits.

> "Neither is wealth necessary merely because it affords the means of subsistence: Without it we should never be able to cultivate and improve the higher and nobler faculties. Where wealth has not been amassed, every one being constantly occupied in providing for his immediate wants has no time left for the culture of the mind; and the views, sentiments, and feelings of the people become alike contracted, selfish, and illiberal. The possession of a decent competence, or the ability to indulge in other pursuits than those that directly tend to satisfy our animal wants and desires, is necessary to soften the selfish passions, to improve the moral and intellectual character, and to insure any considerable proficiency in liberal studies and pursuits. The

[25] Lefebvre, 2014

acquisition of wealth is, in fact, quite indispensable to the advancement of society in civilization and refinement. Without the tranquillity and leisure afforded by the possession of accumulated riches those speculative and elegant studies, which expand and enlarge our views, purify our taste, and lift us higher in the scale of being, could not be successfully prosecuted."[26]

But this time never comes and people continue to work as the predominant notion in the current society is to acquire more and more wealth simply for its own sake. The real question is why this subject occupies such a high position for a citizen in this society than that of the pursuit of science, although a positive effect of mind transformation would occur with the sciences, which would not be the case with work[27] and the acquisition of financial means[28]. Truly, here again the prevailing ideology that has taken hold in our world is revealed[29].

> „The aim of the 'critique of ideology', the analysis of an ideological edifice, is to extract this symptomal kernel which the official, public ideological text simultaneously disavows and needs for its undisturbed functioning."[30]

When money is preferred before everything else – the intellectual hobbies and the further exploration of our intellect stands at the back. What a pity, when so much more about the world can be discovered, if one would take the time to study science and engineering.

---

[26] Smith, 1776
[27] Necessary labor required to earn yourself a living.
[28] Obsession with gaining more and more money.
[29] The prevailing ideology that is capitalism.
[30] Žižek, 2020

"There is no doubt about the actual reason why I dropped out of the technological system. It reduces people to gears of a giant machine. It takes away our autonomy and our freedom. My motives are certainly not altruistic. I just don't like living in this damn system."[31]

Returning to the topic of why there is a lack of interest in science and engineering, it can be said that people see no need for it because the benefits of taking the time and effort to study those topics are not immediately apparent. Let's be clear: Studying science, mathematics, and engineering is hard. They are a challenge to many students, and the lack of adequate preparation for its demands can lead to feelings of inadequacy and frustration. Students are often exposed to mathematics without any prior warning, and the complexity of the subject can be overwhelming, leading them to conclude that they are less intelligent than their peers. This is further exacerbated when they discover that their parents also struggled with the subject, which can lead them to assume that math is an innate talent that runs in the family, and that those who excel in it are prodigies. All these experiences do not give a good feeling at all when dealing with such challenges. This further results in a reluctance to engage with complex subjects and can even dissuade students from pursuing careers in complex fields – careers that require a degree in engineering or advanced mathematics. How ironic that such careers are also very well paid.

"Man is an infinitely complex being and his knowledge entails a multitude of aspects, investigations, techniques"[32]

---

[31] Kaczynski, 2018
[32] Lefebvre, 2014

In my opinion, the current educational system does not provide students with the necessary prerequisites for fostering an interest in science. Students might be more motivated to study science if they could see its real-world applications in their daily lives. They may not fully comprehend the significance of science, nor appreciate its practical applications. Instead, they may view science as a subject that requires much effort and dedication solely for the purpose of achieving good grades. What a boring objective. As a result, of course, many students fail to develop an interest in science, leading to a lack of motivation in pursuing further education in this field. By exposing students to the beauty of science, educators can inspire students to develop an interest in this field. In this regard, science teachers must go beyond teaching science as a subject and demonstrate its practical applications in everyday life. Not only that, they should also show the highest career one could achieve as a scientist or engineer – what a motivation that could have been for me at that time!

Let us not forget the importance of parental influence in shaping a child's career aspirations – especially in academic fields. There is a huge potential advantage that children with academic parents may have over their peers. I have experienced it first hand: A friend of mine studied physics most probably because his father also works as a physicist. Even if the motivation was not to follow in his father's footsteps, spending most of his childhood with a physicist might have had that enormous impact. Children of academic parents are more likely to pursue careers in academia due to the influence and exposure they receive from their parents –

this phenomenon is known as academic socialization.[33] My grandfather was a brilliant engineer. But I couldn't exchange ideas about engineering with him back then because I was too young to understand it. Nevertheless, he is an inspiration today and I am sure that he would have had much more influence on me if I had been able to exchange ideas with him in the past. Would you consider studying science as something creative or rather as rigid, rule-bound, and lacking in creativity? That was my perception before studying it at university. In reality, science is a dynamic and creative field that requires imagination and innovation to make groundbreaking discoveries.

"Being open to using unconventional procedures, chance, and serendipity are needed to foster creativity. Unexpected results can provide new insights into existing problems or clues to solving them."[34]

This perception of science as lacking in creativity and imagination may stem from a misunderstanding of the scientific method – a systematic approach to investigating the natural world. Yet, it is not a rigid set of rules that scientists must follow blindly but rather a flexible framework that allows for creativity and innovation in the pursuit of knowledge. Such is the thing when one does not understand the whole of science before studying it in detail.

In retrospect, all of these things and more could influence why someone has an interest or not in science and engineering. That will be difficult to determine. In any case,

[33] Taylor et al., 2004
[34] Reche & Perfectti, 2020

there is no guarantee that someone will enter this field of study just because certain requirements are met - but it is a start. At least the general conditions can be improved. There are many factors that contribute to an individual's interest or lack thereof – but meeting certain requirements or conditions does not mean that an individual will pursue a career in science and engineering either. Some of the potential influences include academic exposure, personal interests, societal pressures, and cultural and socioeconomic factors. It is evident that a shortage of interest in scientific fields leads to certain issues that warrant further discussion, as elaborated below. These issues or problems are relevant for each individual as well as for the totality of all individuals, i.e. our society. While the concerns of society are undoubtedly important, this discussion will primarily focus on the problems encountered by individuals as I would like to awaken the motivation in every person to deal with complex sciences. It is incumbent upon individuals and institutions to invest in scientific education, work together, and promote a culture of innovation and collaboration, to improve scientific education and scientific literacy. Scientific progress depends on individuals who possess the knowledge, skills, and motivation necessary to engage with scientific research. A lack of interest thereof to engage with sciences leads to individuals being unable to appreciate its splendor fully. This can further impede an individual's ability to develop a deeper understanding of science and their far reaching implications.

Individuals must recognize that science is a multidisciplinary field that requires and promotes a diverse range of skills and knowledge. Its benefits reach out farther than the individuals current employment, and the implications are much more important than just solving current life's problems. It is important to promote a culture of scientific inquiry and go beyond simplicity, so upcoming problems that our society faces can be dealt with more easily. It can be said that one potential solution to this problem is to promote scientific curiosity and encourage individuals to explore scientific concepts from an early age. However, I would like to focus on what an individual can do *here* and *now* to improve their *interest* in science and engineering. First, let's look at the *benefits* of such an endeavor.

Why You Should Focus On Science

"What is my answer to this dilemma? One answer is you must concentrate on fundamentals, at least what you think at the time are fundamentals, and also develop the ability to learn new fields of knowledge when they arise so you will not be left behind, as so many good engineers are in the long run."[35]

Acquiring a robust understanding of scientific fields and engaging with complex topics enables one to approach information with a more critical eye. An individual who possesses a deeper knowledge of complex topics – in general, that is, understanding how complex concepts are usually *conditioned*[36] – is less likely to be swayed by simplistic explanations or oversimplifications, as they recognize the intricacies and nuances that underpin complex subjects. This heightened understanding of the *complex*[37] also allows for a greater appreciation of their inherent complexity, revealing the multifaceted nature of scientific fields and the interplay of their various components.

The importance of learning complex concepts is itself a topic of great interest to scholars across disciplines[38]. You probably already have an idea why this is the case, but I would like to discuss it in this chapter. While the advantages of such

[35] Hamming, 2022

[36] I will sometimes refer to this condition in general as (the condition of) the *complex*.

[37] How complex concepts are conditioned.

[38] Murphy, 1988 | Keller et al, 1991

an in-depth understanding may not be immediately apparent, they are far-reaching and multifaceted for those who take the time to study them. By delving deeper into complex concepts, readers will also acquire a deeper understanding of the subject matter that extends far beyond the specific concept under consideration. Through rigorous analysis, this book seeks to demonstrate the importance of moving beyond simplistic representations of scientific concepts – not only for science, for the understanding of our universe and life itself.

Achieving a deep understanding of a subject is an essential aspect of learning. If one only scratches on the surface and and be content with superficial knowledge, it will prevent them from fully grasping the limits of their understanding – this makes it difficult to find the motivation to understand more complex topics in the future. If one only learns to the point where they can understand the material without difficulty and gives up easily when faced with challenges, they are unlikely to develop a deep and nuanced understanding of the subject matter. This can be particularly problematic when attempting to engage with other sciences, as the gaps in one's knowledge may prevent them from fully grasping the intricacies of the new topic.

Not only that, but by failing to develop a perseverance when dealing with something more complex, giving up as soon as the topic becomes more difficult, learning only to the point where they can understand the material without difficulty, and then failing to engage with the complexities and challenges that the subject presents, one misses out on the valuable opportunities that the science hold. This pursuit of in-depth knowledge is particularly important in academic research, where the aim is to contribute to existing knowledge and push the boundaries of our understanding. Conducting research that is merely based on surface-level information may result in incorrect or incomplete findings, hence why it is important to go in-depth.

I want to change your perception of learning the most complex concepts. I want you to see that it makes sense to learn things that we do not need immediately. Especially if the concept is difficult to understand, it makes sense to take on the effort and learn it nevertheless. Constantly exposing yourself to new concepts, learning more, learning more difficult things, that's what I'm about here. In the end you will have a lifetime of homework, find meaning, enriched your life and enjoy it more than ever. In this regard, the importance of learning new complex concepts cannot be overemphasized as it is a crucial aspect of intellectual growth:

- *Different worldview*: One advantage of learning new, complex concepts is that it changes the way we look at the world. As we learn new ideas, we create fresh mental models that help us comprehend occurrences that were once beyond our comprehension. For instance, when we study the fundamentals of Physics, we start to comprehend how the universe functions, from the tiniest subatomic particles to the grandest cosmic architecture. In this sense, novel ideas open up previously closed vistas of knowledge and comprehension. It provides access to even more information and comprehension.

- *Enhanced intellectual skills*: It enhances our memory and cognitive flexibility, which makes learning new things simpler. Our brains create new neural connections as a result of learning something new, strengthening our cognitive infrastructure and making it simpler to recall and use what we have learnt. This is why acquiring new ideas can be viewed as a type of mental workout that develops our mental faculties.

- *Improved Confidence*: We have the courage to take on even more tough concepts once we have mastered a difficult subject that previously appeared unachievable. This can start a positive learning cycle whereby each new idea we grasp offers up fresh perspectives on knowledge and comprehension, inspiring further growth and learning.

- *Better problem solving skills:* When faced with a difficult problem, we frequently feel overburdened and unsure of what to do next. However, if we have a thorough grasp of the pertinent ideas, we can use them to divide the issue into more manageable parts. Finding practical solutions and identifying the main problems may become simpler as a result.

- *More Learning strategies*: Learning difficult concepts can also help us become better learners. To properly absorb a new topic that is difficult for us to grasp, we must employ a variety of learning techniques. This may entail reading, writing, sharing the idea with others, and implementing it in real-world situations. As we hone these abilities, we improve as learners in general and are better equipped to take in and remember new information.

It is essential to note that regardless of how many reasons are presented, personal experience will always differ – but in order to obtain them, you have to *start*. And above all, you have to have a genuine interest in the subject matter, as it will require a significant investment of time and effort to learn it. It is through this passion that we are able to persevere through challenging concepts and coursework, and ultimately achieve a deeper understanding of the material. In the pursuit of learning complex topics, it is essential to prioritize effective time management, develop analytical thinking skills, and most of all, a passion for what you are trying to do with them.

After completing this book, all this will no longer be an issue, because you will have understood how complex concepts can be learned. I hope that I can provide you with the right methods so that you are appropriately equipped for your endeavor. Let's dive right in and let us go *beyond simplicity*…

> "One of the greatest challenges of the human intellect is the discovery of a unified theory for the four fundamental forces in nature based on first principles in physics and rigorous mathematics. For many years, I have been fascinated by this challenge."[39]

Science is a field that requires deep analytical thinking, creativity, and problem-solving abilities. People who possess these traits are well-suited to engage with scientific research

---

[39] Zeidler, 2004

and contribute to scientific progress. A genius[40] is often drawn to science for this reason: they love the challenge!

> "As a child, Leibniz had little inclination to play, preferring instead to read history, poetry and literature. [...] Thus he began to teach himself to read Latin, building on the instruction he received in school."[41]

My own motivation to deal with complex topics came from the motivation to become a *genius*[42]. Besides all the reasons why it is useful to learn more, I hope to motivate people to deal with complex topics through my personal story, which I will outline below.

The catalyst for my transformative journey can be traced back to a specific publication: a book that would alter the course of my life. The title in question, *Speed Reading Genius*, was one that I initially approached with casual curiosity, never anticipating the profound impact it would have on my personal and professional trajectory. Through the creation of this book, I have developed a novel theory that suggests it is possible to attain genius-level abilities by consistently producing *output*[43].

During the process of writing the book mentioned, I have integrated this theory into my own practices, continually striving to enhance my *output* by generating superior writing

---

[40] This term is used here for people that meet the definition to *genius* according to my book *Speed Reading Genius*.

[41] Aiton, 1985

[42] A very loose term one can hardly define and which I tried to define in *Speed Reading Genius*. It is also the reason why I therefore employed the term *Genius Mastermind* for myself – to remind me why I took on this challenge.

[43] *Output* is a work of genius that is also recognized as such. Before that, however, it is above all only a work – a book, a scientific paper, etc. It is through this work that the genius realizes itself – a work of genius. Thus, it is through this work of genius, through which the genius therefore realizes itself. Tthe genius therefore can also be called a work-of-genius-being.

and sharing innovative ideas through publication. I have found myself not only consuming more information, but also retaining it more effectively. The capacity to rapidly sift through large quantities of data, while extracting and synthesizing key insights, facilitated a more sophisticated approach to problem-solving. Additionally, my newfound proficiency in knowledge acquisition opened up new doors in my academic pursuits, allowing me to engage with complex texts in a more nuanced and comprehensive manner.

And so I started studying in a new field, namely business informatics[44]. While undoubtedly intriguing, this field was characterized by its inherent complexity, and as such, proved to be quite challenging. Despite the difficulty, however, I found great personal satisfaction in the pursuit of knowledge within this discipline.

> "It is an extremely useful thing to have knowledge of the true origins of memorable discoveries, especially those that have been found not by accident but by dint of meditation. It is not so much that thereby history may attribute to each man his own discoveries and that others should be encouraged to earn like commendation, as that the art of making discoveries should be extended by considering noteworthy examples of it."[45]

The coursework was complex and demanding, requiring me to utilize critical thinking and problem-solving skills on a regular basis. Despite the challenges, or maybe because of them, I was determined to succeed in my studies, and the rewards were significant. I spent countless hours pouring over

---

[44] A multidisciplinary field that combines principles of business management with the use of information technology to optimize organizational processes and decision-making.

[45] More, 1934

textbooks, practicing coding exercises, and collaborating with classmates to better understand the subject matter. But over time, I found that my hard work was paying off.

I learned more about programming in several languages and developed a significant appreciation for mathematics. I have found that it provides a unique way of viewing the world around us: Mathematics can be used to model and predict natural phenomena, optimize complex systems, and develop new technologies. This has led to countless breakthroughs in fields such as engineering, physics, and computer science.

Now I would recommend everyone to study mathematics, because it is also a beautiful subject in its own right. The elegance and symmetry of mathematical proofs and the intricate patterns found in fractals and other mathematical structures are truly awe-inspiring. I personally found my deep love for mathematics when I encountered topology. Topology is a field that is concerned with the study of objects that are defined by their properties, rather than their specific details. It can be used to study the properties of objects in three-dimensional space, as well as in higher dimensions. It can be used to study the structure of networks and graphs, as well as the geometry of surfaces and manifolds. But why is all of that important for you as a reader? By setting myself the goal of becoming a genius, I have found the motivation to deal with complex issues. By setting oneself the objective of becoming a genius, individuals can motivate themselves to tackle complex concepts that may seem daunting at first. This approach not only enhances the problem-solving skills but also fosters a growth mindset[46] that allows for continuous improvement.

---

[46] The growth mindset is a motivation theory proposed by Dr. Carol Dweck from Stanford University. It means that your ability to learn is not fixed as it can change with your effort.

You see, it all started with my own endeavor of becoming a genius and it led me to study informatics, mathematics, and now engineering.

Set your mind on becoming a certain person with certain characteristics. Do not focus on learning mathematics or physics, but try to become a person who uses these subjects in your every day life. For example, instead of simply focusing on learning mathematics or physics, you can strive to become someone who uses these subjects regularly. Focusing solely on learning a specific subject, such as mathematics or physics, can limit your growth and development as an individual. While knowledge in these subjects is undoubtedly valuable, it is equally important to develop the problem-solving skills, critical thinking abilities, and creativity necessary to apply this knowledge in real-world situations. By striving to become someone who uses these subjects regularly, you are not only learning the subject matter but also developing the skills and mindset necessary to apply this knowledge. This means developing a deeper understanding of the subject matter and applying it in real-life situations. By doing so, you are not just learning the subject matter but are also developing problem-solving skills and critical thinking abilities that can be applied to other areas of your life.

The impact of *Speed Reading Genius* on my intellectual and professional development cannot be overstated. I was able to elevate my level of discourse, engage with a broader range of perspectives, and ultimately cultivate a more refined and sophisticated approach to knowledge acquisition. Through the process of writing about the nature of genius and its development, I gained insights into the cognitive and behavioral patterns that define it, thereby enhancing my own intellectual abilities and knowledge and I developed a skillset that enabled me to digest information at an unprecedented pace.

Furthermore, as I delved deeper into the field of informatics, I discovered a newfound appreciation and love for the intricacies of mathematics and engineering. A love, that I want to share with you in this book. While my experience in informatics led me to develop a passion for mathematics and engineering, I recognize that the journey of discovering one's intellectual passions is a highly personal and individual process. As such, I can only offer insights into my own experiences and hope that they may serve as a guide for others to embark on their own unique path of discovery.

By envisioning my future self, I drew upon the inspiration needed to overcome obstacles[47] and continue down the path of learning complex subjects. It is essential to think long-term and consider where one wants to be in a decade or more. Define a label for your future self and then attach that same label to yourself now. Then read the necessary books your future self would read. While reading, remind yourself of the person you wish to become.

We should first develop self-awareness that allows us to comprehend what works best for us and what does not. We can then seek out guidance that is tailored to our unique aspirations and circumstances. One of the ways to do this is to develop our own image of ourselves – a concept which I call a *vision* in this book. The vision answers the following question: *Who* do you want to be?

---

[47] Herman, 2019

Developing a personal *vision* is an important step towards personal growth and success, as it provides a sense of direction, purpose, and motivation. However, developing a personal vision is not an easy task, as it requires introspection, self-awareness, and a deep understanding of one's values and aspirations. To define your own vision, you could ask yourself the following questions:

- What do I want to achieve?
- What do I care about the most?
- What motivates me to take action?
- What legacy do I want to leave behind?

As success means different things for different people, it is important to first define what *success* means to us. While certain individuals online may have achieved what they consider success, their definition of success may not be applicable or even desirable for us. Social media is often not a good source of inspiration to find our own meaning, yet many people turn online and try to find exactly that. In this chapter, I want to show you how you can create a motivatonal vision of yourself and then pursue it to learn more, more complex topics, and *go beyond simplicity…*

"You are not tied to a particular position; your loyalty is not to a career or a company. You are committed to your Life's Task, to giving it full expression. It is up to you to find it and guide it correctly. It is not up to others to protect or help you. You are on your own".[48]

[48] Greene, 2013

"to see the world as independent and separate from the Self, is an illusion."[49]

The process of reading is intimately intertwined with the formation of one's consciousness and selfhood, as it enables the construction of a unique, individualized perspective. That is to say, you will *become* what you read.

> "Change is inevitable, particularly in such a revolutionary moment as ours. Since you are on your own, it is up to you to foresee the changes going on right now in your profession. You must adapt your Life's Task to these circumstances. You do not hold on to past ways of doing things, because that will ensure you will fall behind and suffer for it. You are flexible and always looking to adapt."[50]

This implies that the content that one consumes through reading contributes significantly to the construction of an individual's identity. As readers engage with a text, they draw on their prior knowledge and experiences to make sense of the content presented to them. In doing so, they actively construct an understanding of the world that is unique to them. This individualized perspective serves as a lens through which readers interpret and interact with the world around them.

> "[...] the nature of Reality, consisting of universal Consciousness, the Self of every being and at every level of existence, the individual human being, and the world of objects, all form a triad of undivided wholeness."[51]

---

[49] Kafatos & Yang, 2016
[50] Greene, 2013
[51] Kafatos & Yang, 2016

In this way, reading is a highly transformational and generative activity that enables people to meaningfully develop their mental and emotional capacities. Indeed, the texts one chooses to engage with possess the power to shape and mold one's character, as the ideas and values contained therein become assimilated into the fabric of one's being. This means that what we read, we become – therefore we would need to first define *who* we want to become in order to choose the appropriate books that we want to read.

Another strategy for developing a personal vision is to identify your role models. Role models are individuals who inspire you and embody the qualities and characteristics that you admire and want to emulate. To identify your role models, you can ask yourself the following questions:

- Who do I admire and respect the most?
- What qualities and characteristics do they possess?
- What challenges and obstacles did they overcome?
- What strategies and techniques did they use to succeed?

Once you have defined your values and aspirations and identified your role models, you can create a vision statement that summarizes your personal *vision*. A vision statement is a concise and inspiring statement that describes your ideal future self and the impact that you want to make on the world. A very elaborated way of doing this, is creating a *biography* of your future self – written by you in the third person perspective.[52] By doing this, you gain clarity on your personal vision and can communicate it to others as well. A personal vision is not a one-time event, but rather a continuous process of self-reflection.

Now I would like to talk about my own personal vision. What I have taken as inspiration is a mixture of various universal geniuses, Renaissance Men and other intellectuals. In the end, this results in an overall picture that should represent my future in the form of a compact vision, that can be described in a simple label: *Genius Mastermind*[53]. The Genius Mastermind is not merely a concept but a self-termed identity that I aspire to embody. It represents the culmination of my personal vision, a fusion of the attributes of individuals who were well-versed in various disciplines, including art, music, literature, philosophy, and science. What has changed since I have defined this vision for myself?

It has helped me tremendously in many areas of my life. I started to write more books and in the pursuit of producing more complex and even better books, I have been extensively learning and expanding my knowledge. To improve my pursuit, I have enrolled in university courses and plan to continue doing so in the future. It is the combination of learning and writing books that has helped me to understand the world better.

"Reading and writing were therefore inseparable activities. They belonged to a continuous effort to make sense of things, for the world was full of signs: you could read your way through it; and by keeping

---

[52] Write a biography about yourself as if someone else would write about you. As good biographies about people who have succeeded in life are usually not auto-biographical. What would this person highlight? Why is it important?

[53] The benefits of the creation of an alter-ego have been shown in Herman (2019).

an account of your readings, you made a book of your own, one stamped with your personality."[54]

My new passion for complex topics has driven me to specialize in this area, that is, the pursuit of knowledge in branches of science which are not easy to grasp. Through these efforts, I have gained valuable insights into various more topics, which has enabled me to write better and more informed books. One might underestimate the overall benefits of doing so, which is probably why so few people do it. I have further expanded my critical thinking skills, which enabled me to analyze complex concepts and synthesize them into comprehensible and informative pieces – more books. This has in retrospect enabled and also motivated me to focus on and find the courage to pursue even more complex topics.

> „For centuries, artists and intellectuals from Leonardo da Vinci to Virginia Woolf, from John Locke to Octavia Butler, have recorded the ideas they found most interesting in a book they carried around with them"[55]

I have also honed my analytical abilities by conducting thorough research, analyzing data, and evaluating sources to ensure the accuracy and credibility of my work in the last couple of years and I can see the difference of quality of my writing. In short, I can say that taking all this time for my intellectual pursuit was absolutely worth it.

As an aspiring genius mastermind, I find myself awestruck by the vast array of intellectual giants that have preceded me. Their works are not only a testament to their

---

[54] Forte, 2022
[55] Ibid

intellectual prowess but also serve as an inspiration to pursue my own academic endeavors. Reading their works can provide insights into their thought process, approaches, and techniques, which can be inspiring and informative for us to improve our own skills and knowledge. It can also serve as a means to clarify personal goals and aspirations.

By analyzing the thought process of geniuses we can gain a deeper understanding of how to approach complex problems. Someone who wants to be a mathematician may be inspired by the works of mathematicians like Serge Lang and John von Neumann and strive to learn their techniques for solving mathematical problems. When it comes to engineering, reading the books of individuals who have achieved mastery in this field can be a good source of inspiration. It is difficult not to feel humbled in the face of such extraordinary achievements and it sometimes surprises me that not everyone shares the same admiration for these remarkable individuals like I do myself. The magnitude of their accomplishments highlights and reminds us of the importance of pursuing intellectual excellence in one's own work.

> "Here's the thing about playing the long game: at times, it can be lonely, maddening, and unfulfilling. It's worth it in the end; we know that intellectually. But in the moment, it often feels like a complete, humiliating waste of time."[56]

As their ideas have shaped the course of history, influenced public policy, and impacted societal values, the works of these great thinkers serve as a reminder that there is always room

[56] Clark, 2021

for improvement and that our own intellectual pursuit is worth our time. Sometimes, I get stuck again. So I read their books to remind myself of all the great accomplishments of people that lived before me. I am in great awe when I think about all the intellectuals, thinkers, and geniuses and their work: Aristotle to Einstein, these thinkers have left an indelible mark on the intellectual landscape of humanity.

> „You must rely on your image of yourself;
> you must build it every chance you get.“[57]

Without a clear idea of the direction you want to go, you are unlikely to learn more. However, I cannot tell you who or what you should do with your life – you have to find out for yourself. How do you do that? By having a clear idea of who you want to be – by defining a *vision* for yourself. How you see yourself – your *vision* – determines what books (*output*) you will write - because it determines what you will learn – and vice versa. If you already know you will want to write a book on physics, mathematics, or philosophy, it becomes relatively easy to choose your next topic. Defining a vision is the first step in deciding what direction you want to go. If your vision is to be a respected mathematician or physicist one day , then you know that to do that you will have to learn mathematics or physics. But why should we want to become a great mathematician in the first place? It is because of our own understanding of our vision. With a vision, we create an idea of who or what we want to be in the future – and we write it down in detail in our own biography. Let's call this

[57] Maltz, 1978

project the *vision biography*, or take the Polish word *biografia*[58] to distinguish it from all other biographies.

> "Another powerful way to frame your choices is to understand your ideal lifestyle. Where, and how, do you want to live? And what would it look like for you to stand up for that vision?"[59]

By writing a *biografia* of someone who meets just these certain characteristics, you will have more clarity about what your *vision* will be and thus what you will focus on in the subject matter. The process of writing about your future self – who has achieved greatness – is called *prospective writing* and it has been shown that it a positive impact on one's life.[60] Describe yourself not in the actual current state, but in the state of the person you would want to be in the future. For this to work, it is advisable to read other biographies of those very people who are close to the very vision you want to achieve yourself. If we tell ourselves we want to be a great mathematician, we get the courage and motivation to deal with mathematics. You cannot be a great mathematician without having accomplished – that is, written, created, developed – mathematical texts (*output*).

> "There's no magic formula—not for ourselves, and not for the people around us. We won't make ourselves more creative and productive by copying other people's habits, even the habits of geniuses; we must know our own nature, and what habits serve us best."[61]

---

[58] Polish words are used for vision terms in this book. I would prefer Chinese words, but these are rather difficult to decipher.

[59] Clark, 2021

[60] Roepke et al., 2017

[61] Rubin, 2015

It is by setting ourselves *challenges* according to our *vision* that we will find the motivation to engage with complex scientific topics. Let us look at what such *challenges* are and how to set them.

"When you are born, your neocortex knows almost nothing. It doesn't know any words, what buildings are like, how to use a computer, or what a door is and how it moves on hinges. It has to learn countless things."[62]

[62] Hawkins, 2021

Embracing Complexity

Strategies for Studying Complex Concepts

"Learning is not attained by chance, it must be sought for
with ardor and attended to with diligence."
Abigail Adams

Intellectual Challenges

If you would know how your brain reacts to an intellectual challenge, you would probably engage more in such challenges. Let us look at the facts:

"Although the human brain is exceptional in size and information processing capabilities, it is similar to other mammals with regard to the factors that promote its optimal performance. Three such factors are the challenges of physical exercise, food deprivation/fasting, and social/intellectual engagement."[63]

However, we must distinguish between challenges that come our way and those we set ourself. I will first discuss the challenges that will undoubtedly be encountered, i.e. a challenge not set by ourself. You could also consider them as the *bad* challenges, but you will realize that every challenge you face will have a positive effect on you in the long run. First, I would like to focus on the most difficult challenges that I had faced when learning something complex. Please be aware that everyone faces different challenges and the difficulty of dealing with each of them is a subjective matter.

One of the primary challenges of learning complex concepts is their inherent complexity. The complexity of a concept can make it difficult to understand and can be a significant obstacle that can hinder the understanding and

---

[63] Mattson, 2015

retention of key ideas. Complex concepts are often abstract, multifaceted, and difficult to grasp, making them challenging to understand and remember.

"If we are to have a true knowledge of an object we must look at and experience all its facets, its connexions and "mediacies". That is something we cannot ever hope to achieve completely, but the rule of comprehensiveness is a safeguard against mistakes and rigidity.' It is instructive to see here how an abstract philosophical category, deepened by epistemological provisos governing its application, serves directly as an imperative to correct practice."[64]

A key strategy to overcome this challenge is to break down the concept into smaller, more manageable parts. This approach can help learners to better understand the different elements of the concept and how they relate to each other. This interconnectedness can make it challenging to identify the most important components of the concept and how they relate to each other. There are also a lot of other challenges that come with such relations:

"These powers are themselves located in a network of relations between exploiters and exploited involving those clashes between different class practices that make up the class struggle: in short, they are inserted in a system of inter-class relations."[65]

Complex *concepts* are often built on a foundation of underlying principles and assumptions that the learner has to know before dealing with such a concept – it is almost like a complex concept is consisting of many different concepts in a whole system. Such is the complexity of concepts and of the

---

[64] Lukács, 2009
[65] Poulantzas, 2014

systems such concepts consist of. We are completely at the mercy of such systems.

> "[…] we are determined by systems and forces completely outside our capacity to affect them."[66]

The volume of information required to understand such a system can be overwhelming, leading to cognitive overload and difficulty retaining information. But without this prior knowledge, learners may struggle to understand the context and significance of the concept itself. Therefore, it is essential to identify and address any knowledge gaps before attempting to learn a complex concept fully. Most of the time, it is not possible to know the relevant concepts necessary to understand the whole system of concepts that the complex concept entails.

One of the most important habits to cultivate in this regard is the habit of completing something every day. Even if the project is small and seemingly insignificant, the act of finishing it can help to build momentum and to keep one's focus on the larger goal. By breaking down larger projects into smaller, more manageable pieces, it is possible to maintain a steady stream of progress and to avoid becoming overwhelmed or discouraged. It is important to embrace challenges as opportunities for growth and learning, rather than being daunted by the prospect of failure or criticism, and approach each challenge with a sense of curiosity and excitement. We have to view obstacles as opportunities to develop new skills and knowledge. In the realm of creativity

[66] Dean, 2019

and intellectual pursuits, it is a rare phenomenon for someone to achieve lasting fame and recognition for a single achievement. Geniuses have faced – and overcome – a number of challenges, ultimately making them who they were at that time and for what they are remembered today. They have produced a significant volume of works[67] in order to be acknowledged as such. Very few geniuses demonstrate exceptional abilities from an early age, and it is likely that their brilliance develops over a period of time through a combination of learning, practice, and experimentation. Challenges have served their purpose by helping geniuses achieve just that, i.e. by overcoming them. While it is difficult to determine with certainty whether someone is a genius before or after they have completed such challenges – and amassed a significant body of work while doing so – it is clear that the development of genius takes time.

> "For, as I said, there is no genius without a productive force, and furthermore, it does not depend at all on the business, the art and the profession that one pursues: it is all the same. Whether one proves ingenious in science like Oken and Humboldt, or in war and state administration like Frederick, Peter the Great and Napoleon, or whether one makes a song like Béranger, it is all the same and depends only on whether the thought, the aperçu, the deed is alive and able to live on."[68]

---

[67] I usually consider such works as *output*.

[68] Translation of Eckermann (1885): *"Denn, wie gesagt, es gibt kein Genie ohne produktiv fortwirkende Kraft, und ferner, es kommt dabei gar nicht auf das Geschäft, die Kunst und das Metier an, das einer treibt: es ist alles dasselbige. Ob einer sich in der Wissenschaft genial erweiset wie Oken und Humboldt, oder im Krieg und der Staatsverwaltung wie Friedrich, Peter der Große und Napoleon, oder ob einer ein Lied macht wie Béranger, es ist alles gleich und kommt bloß darauf an, ob der Gedanke, das Aperçu, die Tat lebendig sei und fortzuleben vermöge."*

Indeed, it is important to recognize that everything we do, no matter how seemingly effortless or innate, is the result of a learning process. Nothing is truly innate – even the most talented individuals must dedicate significant effort to refining their skills and honing their craft. Whoever creates something new is *creative*, whereby what is created is called a *product* of his or her creativity. Those who do this again and again are being *productive*. I would like to explain here, that just writing a lot of books and papers is of course not enough to be genius. It is the combination of creativity and the means to repeat the process over and over again. Getting more ideas depends on your ability to write and publish more often, but the ability to find the one crucial idea that should be turned into a research paper, i.e. seeing an interesting twist on an existing topic or already imagining a new discovery, would be dependent on the so called *Q-Factor*[69] – i.e. your ability to turn ideas into great research papers. When I am talking about *productivity* in this book and in regards to geniuses, I do not just mean the publication of a lot of books, but rather the publication of scientific writings with a new twist, idea, or discovery. Hence, such productivity concludes into a lot of *output*. The question now is: How were geniuses productive? By setting *challenges* for themselves. The distinguishing characteristic of successful geniuses is their ability to set their own challenges with which they achieved better learning outcomes and significant *output* as well. This is evident in their various accomplishments, as they were driven by their innate curiosity to explore new knowledge and territories. In light of this, it is essential to cultivate an interest in and

---

[69] Barabási, 2023

curiosity for novel ideas, something that is typical of geniuses. These individuals focused on one *challenge* at a time, diligently working on it until its completion before moving on to the next. Therefore, challenges in this sense are self-imposed, with the objective of discovering new knowledge or creating something new. How then can one set such challenges for oneself?

Gottfried Wilhelm Leibniz designed plans for a submarine, not because anyone told him to do so, but because he set the *challenge* for himself. Similarly, Leonardo da Vinci created designs for airplanes, and Athanasius Kircher studied the plague, all without any external prompting, i.e. people paying them money to explore these areas. As students, we encounter numerous challenges in college, such as tests, theses, and projects, which we undertake and complete at specific times. However, what about our free time? Why not incorporate these challenges into our daily lives as well? Geniuses conducted research mostly in their spare time – it was *unpaid* work. Moreover, their successes often led to their being entrusted with crucial responsibilities – which then resulted in more *paid* work. Although they may not have been aware of their ability to meet these tasks, they accepted the challenges anyway.

> „If somebody offers you an amazing opportunity, but you are not sure you can do it – say yes, then learn how to do it later." – Richard Branson

Setting a challenge on your own is a way of gamifying the learning process – leading to better motivation in learning and better outcomes in the long run[70] – also known as *Challenge-Based Learning* (CBL)[71]. Some benefits of *Challenge-Based Learning* were the following, among others[72]:

- Improved creative problem-solving skills
- Increased curiosity in solving problems
- Viewing problems from different perspectives
- Generating more ideas to solve problems
- Increased willingness to take risks in problem-solving
- Improved resilience when working on difficult tasks

These newly gained skills can subsequently be applied to various other domains of life. One of the most significant benefits of a *Challenge-Based Learning* approach is the improvement in creative problem-solving skills.

> "Challenge-Based Learning, or CBL, is an active methodology which focuses on the development of problem-solving skills through challenges or tasks. This methodology encourages students to work in groups, develop their critical thinking skills, and take an active role in their learning. The objective? To develop the knowledge, skills, and confidence needed to provide a solution to the challenge that's been set."[73]

By dealing with problems from the real world, a challenge encourages us to think deeply and creatively about possible solutions – which then leads to such solutions and also fosters our creative thinking. If we are building actively, we are creating new possible solutions.

[70] Legaki et al., 2020

[71] López-Fernández et al., 2020 | Make sure to explore the difference between *Challenge-Based Learning*, *Project-Based-Learning*, and *Research-Based Learning*.

[72] Yang et al. 2018

[73] Martín, 2023

"Maker activities are student-centered and engage students in innovative design and creation processes. [...] They provide opportunities for students to experience "learning by doing" and solve real-world problems."[74]

In a creative challenge, we can play, make mistakes, and learn. It is about having fun and enjoying the learning process. It is absolutely necessary to engage with as much material as possible in order to internalize the way of thinking that leads to new insights, which in turn lead to creative outcomes. Look for things you don't understand today because they might give you ideas for what you need to learn tomorrow – when we encounter concepts or ideas that challenge our existing understanding, it indicates that we are engaging in critical thinking and questioning, which can lead to new discoveries and advancements. Understanding them and combining them, that is called progress.

„When you're able to think about all of these disciplines together, you kind of think differently and can dream of much crazier things and how they might work. I think that's really an important thing for the world. That's how we make progress."[75]

If something is difficult for us and we have to make an effort, it means we are learning something new. The easier it is for us, it does not mean that we are a great genius, but that we are not yet challenged enough. We should challenge ourselves and work together on understanding complex theory. Through collaboration with our peers, we are exposed to a wide range

[74] Weng et al., 2022
[75] Vance, 2015

of perspectives and potential solutions – solutions that can be successfully established through strategic planning and implementation – fostering creativity and innovation. Bringing people together to examine a problem from all sides and using that knowledge and insight to collectively build a solution is the essence of creative teamwork, i.e. people that work together outside of capitalist exploitation that the ideology of teamwork is usually attributed to.

"A sizable Marxist-inspired literature critiques corporate efforts to use participation and teamwork ideologies to undermine worker solidarity and union organization."[76]

This diversity of perspectives leads to a richer understanding of the problem and the potential solutions. Students also become invested in finding a solution to the problem presented to them, driving their motivation and interest in the subject matter. It not only leads to a more comprehensive set of ideas but also encourages students to be more creative in their thinking.

"Marx argued that it was work, or consciously transforming nature, which distinguished humans from animals, and it was through the creative process of work that man came to recognise himself as human."[77]

By encouraging ourselves to experiment with different approaches and solutions, challenge-based learning creates a safe space for failure and iteration. Through this process, we

[76] Adler, 2009
[77] Thompson, 2022

can develop a growth mindset, understanding that failures and mistakes are opportunities for learning and improvement.

> "[…] the influence of a single idealist thinker could extend to several different Marxist theorists. Bachelard, for example, not only inspired Althusser: he was also admired by Lefebvre, Sartre and Marcuse, who drew quite other lessons from his work."[78]

This way, we are able to develop our own interests and curiosities, and pursue our passions with relentless dedication. It is also important to develop a sense of discipline and commitment, in order to stay focused and motivated over the long term.

In *Challenge-Based Learning*, students concentrated their efforts on a single challenge, rather than spreading themselves too thin by attempting to tackle multiple projects simultaneously. By doing so, they were able to develop a deep understanding of the subject matter at hand, and to approach problems from a unique and innovative perspective.[79] In order to set challenges for yourself, it is crucial to cultivate a strong interest and curiosity in the subject matter. While many of the challenges faced by geniuses throughout history were scientific in nature, there is no reason why the same principles cannot be applied to other areas of your life as well. All you have to have is a willingness to explore new ideas and to approach problems with an open mind, rather than relying on preconceived notions or assumptions.

Whether it is writing a novel, learning a new language, or mastering a musical instrument, the key is to set specific goals

---

[78] Anderson, 1976
[79] Yang et al. 2018

and to work towards them with dedication and determination. Everything can be learned, and anyone can choose to live a more productive and creative life. Even the most gifted individuals must work tirelessly to achieve their goals and realize their full potential. It is not that everything becomes easier the better you become, it is rather that you can now set yourself more difficult challenges – and achieve a greater outcome along the way.

> "One of Lenin's most characteristic and creative traits was that he never ceased to learn theoretically from reality, while remaining ever equally ready for action. This determines one of the most striking and apparently paradoxical attributes of his theoretical style: he never saw his lessons from reality as closed, but what he had already learned from it was so organized and directed in him that action was possible at any given moment."[80]

Throughout history, geniuses have been known for their ability to discover new ideas and concepts, and to push the boundaries of what is possible[81]. What sets them apart from others is their willingness to set their own challenges and goals, rather than relying on external sources to provide them. In the following chapter, I will delve into the concept of *challenges* and explore the process of setting one for ourselves. To illuminate this subject, we will draw inspiration from the remarkable achievements of Gottfried Wilhelm Leibniz, who actively pursued various *challenges* throughout his life, often during his leisure time. By examining Leibniz's endeavors, I aim to provide readers with a clearer understanding of what

---

[80] Lukács, 2009

[81] There are sciences that need to be explored further. Just because the meaningfulness is not understood today, the development of a new discovery can still be of importance later in the future.

constitutes a challenge, how to establish one, and ultimately, how they can embark on their own personal challenges.

"The point was not to pursue philosophy as a purely theoretical exercise, withdrawn from the world in splendid isolation; nor was the intention to use the new philosophies to sweep aside the intellectual rubbish accumulated over the centuries, as Bacon and Descartes had proposed. Rather, if employed as a practical means of bettering the human condition and resolving the national and international, political and confessional problems which plagued the Empire, the certainties which the new science seemed capable of delivering might be used to reform, harmonise, bolster, and defend all that was best in the old theological, ecclesiastical, and political order."[82]

Leibniz made significant contributions to various fields, including mathematics, philosophy, and engineering. In particular, he developed the integral and differential calculus, which he shared with Sir Isaac Newton. It is said that Gottfried Wilhelm Leibniz has set himself many tasks and *challenges*, such as:

- Creating plans for a submarine
- Writing a description of a binary system
- Improving the technology of door locks
- Implementing a device for determining wind speed
- Development of the decimal classification
- Development of the endless chain for mining ore extraction
- Advice physicians to take temperature more regularly
- Proof of the unconsciousness of man
- Infinitesimal calculus

[82] Antognazza, 2011

We can see, Leibniz was a discoverer who took pleasure in developing something new. He conducted research in different areas and in different ways, showing that not everything must be completely new, but combining or improving something can also be significant. A *challenge* should motivate us to deal with complex ideas – it should not discourage us. Therefore, set yourself a challenge that you know you are up to, even if it means that you might not be innovative at first. We can see from Leibniz's example how differently such challenges can appear – I would consider these as different *types* of challenges. The question still remains as to *what* a challenge is, *how* it is defined, and *how* one sets one for oneself.

A lot of sources as well as an official website dedicated to the *Challenge Based Learning* approach state a *challenge* as something that appears in the "real-world"[83]. One might think about solving world hunger or curing cancer – probably the most popular challenges today. However, there are also other types of *challenges* such as unsolved problems in mathematics[84], physics[85], or computer science[86]. Solving these challenges will have a *real-world* application and dealing with them is an interesting, challenging, and most importantly, meaningful endeavor, even if you cannot solve these *challenges* right away. For example, reading ten pages a day is not a *challenge* – it is a *habit*. Neither should writing one hundred books be considered a *challenge* – this is a *goal*. Solving a

---

[83] CBL, 2023

[84] https://en.wikipedia.org/wiki/List_of_unsolved_problems_in_mathematics

[85] https://en.wikipedia.org/wiki/List_of_unsolved_problems_in_physics

[86] https://en.wikipedia.org/wiki/List_of_unsolved_problems_in_computer_science

challenge in a timely manner can be considered a *goal* but let us keep focused on the definition of what a *challenge* is by first defining its three characteristics:

1. *Problem in the real world:* There is a real world application of the final solution as the problem or challenge appears in the real world. This problem does not have to be instantly visible to everybody as long as *you* can define what the problem is. Do not go out and ask people what kind of problems they think have – they do not know. Especially when it comes to complex topics – the focus of this book – it will be impossible for the everyday man or woman to know what kind of problems currently exist. You have to discover them yourself[87], create a hypothesis[88], and sketch out your first draft[89], before even defining a clear *challenge*.

2. *Focused and defined scope:* Solving world hunger[90] is a big issue but so is sanitation and healthcare. Understanding which kind of real world problem you want to focus on and what specific part of it you want to solve is one of the first steps to consider when defining your own challenge. Reconsider these big issues again before diving into them. You should rather focus on a very small aspect of the kind of problem you want to tackle. Otherwise you might be overwhelmed with the amount of work that you will have to do to solve it. The topic you choose is often rooted in your own passion and interest,

---

[87] This has a lot to do with academic research.

[88] A starting point for further investigation.

[89] Get a better understanding on where you are and where you want to go.

[90] The immediate objective need not be to effect sweeping global transformations, as such aspirations are not obligatory. Instead, the pursuit of expanding one's individual reservoir of knowledge proves to be a satisfactory objective in its own right.

i.e. it is a good idea to first look at those before thinking of doing something completely different.

3. *Product as output:* What good is it to have all the knowledge available if you do not do anything with it, i.e. if you do not share it with others? As already mentioned previously[91], the final result of the challenge, that is the *product*[92], needs to be something tangible, i.e. a *book* or a *concept*. Learning or studying complex concepts is not enough to consider it a *challenge* – there has to be a *product* of your intellectual endeavor that one can digest, ingest, and think about. What kind of *medium* to use for your challenge is itself a study of great interest. Setting a challenge for yourself can lead to unexpected discoveries and can serve as a stepping stone for future innovations. However, is important to note that not all completed outcomes will have a clear purpose or find practical use. The driving factor of designing a challenge for yourself is *curiosity*. We display a natural interest about the world around us from an early age, aggressively seeking out new experiences and knowledge. For some people this desire for more knowledge goes beyond the usual – this is also referred to as *epistemophilia*.

> "The curious individual is motivated to obtain the missing information to reduce or eliminate the feeling of deprivation."[93]

When it comes to Leibniz, he mostly taught himself.

[91] See *Why Challenges Are Important*

[92] This term not stem from a commercial intention, but to describe a *product* of your *intellectual* and *creative* capabilities.

[93] Loewenstein, 1994

He was interested in learning more[94], and his personal learning endeavors eventually got him very far. Nobody told a Gottfried Wilhelm Leibniz to design a submarine – it just excited him. As all the other geniuses, he did most of his research in his spare time. In a classroom, it is usually the teacher who sets the challenge as well as the expected *outcome*[95] – as your grade would be based on how much of the outcome you have achieved –, but in the real world you set them yourself, just as Leibniz did. That is the beauty of such a challenge, as you have complete freedom over the issues you want to tackle and the way you are going to tackle them. Unconstrained by the guidance of others, Leibniz ventured into uncharted intellectual territories, delving into disciplines ranging from mathematics to philosophy.

> "Leibniz's earliest teachers actively discouraged his independent, exploratory, and therefore disorderly approach to learning and attempted to restrict his reading to the formulas and methods of Comenius's work. Although Leibniz was not an autodidact in the strict sense of the term, and although the formal education he received at school and especially university was not without its genuine stimuli, the pace and trajectory of his earliest intellectual development would clearly be established outside formal educational institutions rather than within them."[96]

Before we can learn anything new, we must free ourselves from ideologies that keep us stuck in our current barriers. To learn *new* concepts, we must first discard *old* concepts i.e. old thinking about concepts. We should discard them when we

---

[94] Is it not a sufficient motivation in itself to be driven by a personal interest in these subjects and to engage with them earnestly?

[95] Not necessarily true as there are many possible outcomes or a real world problem and the role of the teacher is to "find the Solutions with the students" (CBL, 2023)

[96] Antognazza, 2011

recognize their superiority based on informed considerations, rather than clinging to them out of habit. It will be difficult to learn to distinguish the difference.

It is unlikely that Leibniz saw himself as inherently superior to others; rather, he simply approached his challenges in his own unique way. Each of his works, like any other creation, possesses its own distinctive qualities. Leibniz never aspired to design the perfect submarine or devise an unpickable door lock. Instead, he imbued his creations with his experiences, thoughts, and ideas. Descartes and Cavalieri had already developed the first approaches to calculus, but Leibniz further refined the concepts into what we today know as Infinitesimal Calculus – but, at the same time, this was also developed by Newton. This work eventually led to Abraham Robinson's non-contradictory infinitesimal calculus. Leibniz's achievements demonstrate that two people in different parts of the world can still have similar ideas. However, this should not discourage you from developing and writing down your own ideas. Even if they may not seem original at first, your unique perspective and approach can lead to significant contributions in your field. All you have to do is doing it *your* way – it is important to pursue your own path and not strive to emulate the genius of others without first exploring your own personal inclinations. Leibniz did it *his* way – all his experiences, thoughts and ideas went into his *products*.

You can also study, learn, and write in as many different areas as you like – the concept of specialization is made up to make you more suitable for the job market, but that does not mean that this is the thing that you should do your whole life. Take universal geniuses as role models for your challenges: The interests of these geniuses were wide-ranging and there was always a new concept or idea to discover. Georg Wilhelm Friedrich Hegel was particularly interested in the following subjects:[97] *Ancient languages* such as *Greek and Latin, mathematics, theology, politics, economics, history,* and *existential philosophy*. Geniuses – and in particular *universal* geniuses – had the opportunity to choose to study in a number of fields – which in turn increased creativity in other areas. There were many geniuses in the field of *physics*; nevertheless, each of them managed to be enthusiastic about a particularly specialized field. That does not mean that these were the fields they were particularly drawn to because for financial reasons. It happened that they developed a fundamental interest in something and then they wanted to learn more about their specialization. It seems that the childlike curiosity that is inherent to all of us did not pass by at some point in these people, but rather found a new way to express itself on another level – mainly that of *science*[98]. One universal genius whom I admire was Caramuel y Lobkowitz who wrote, for example, in the following fields: *Grammar, poetry, rhetoric, mathematics, astronomy, physics, politics, canon law, logic, metaphysics,* and *theology*. While others were busy solving puzzles in their school days, Lobkowitz proved himself at mathematical problems. He wrote about *probability theory* in connection with the game of dice and tried to solve theological questions mathematically. Today, we can all choose to spend our free time on science or studying foreign languages, mathematics or theology. It becomes scientific – and ingenious – as soon as we *write down* our thoughts and

---

[97] According to various sources which I am too lazy to mention here.

[98] Society attributes the term *genius* to someone who spends their time dealing with scientific topics and creates significant *output* too.

make it clear how these ideas came to be. Not all of Lobkowitz' works have garnered the scientific acclaim that aligns with contemporary standards – a fate shared by the works of Leibniz as well. The fact that most of these works have faded into obscurity in our present era is not inherently tragic. Instead, should we not find satisfaction in the recognition that a selected few of his works continue to be revered as remarkable?

Whether it be metaphysics, ore mining, or even erotica, geniuses expressed their thoughts through writing, commentary, and review. As we delve deeper into the world of understanding the thoughts and ideas of geniuses, we realize that they often gravitated toward similar topics[99]. Remarkably, even when it seemed that every realm had been thoroughly explored, geniuses continued to make new discoveries. More importantly, it was not only the scientific fields they chose that bear similarities, but also their perspectives. The simple takeaway from all of this is: Do what excites you. I cannot provide a comprehensive explanation as to why you should undertake specific subjects – a fascination for a subject can at most be described, but never truly explained. I cannot, for example, convince you to delve into the realm of higher mathematics – only explain why it is important. Explaining the significance of a particular piece of knowledge is not always a straightforward task as the importance of a thing can only be experienced through dedicated and intensive study. The reasons are numerous but too abstract to grasp – especially if you do not know higher mathematics. In fact, any reason I would give you is just my personal motivation and it

---

[99] Science and science related.

has nothing to do with you. I study these topics because they excite me, but that is not enough of a reason for many. One might as well learn Spanish or French, because it is easier and they enjoy it much more. But as mentioned before, there is a significant difference between joy and *obsession*. If you indeed have more accessible alternatives available to you, it is reasonable, based on your own perspective, to choose not to pursue mathematics – then study something else.

In examining Leibniz's inventions, it becomes evident that while some of these inventions were meaningful and significant, others were not. However, it is worth exploring new ideas and completing them to the best of your ability. Let's look at them in more detail and get a feeling on what these *challenges* entailed:

Leibniz developed a dual system, which he saw as a symbol of the Christian faith. Although the concept of a dual system had already been introduced in India in the 3rd century BC, Leibniz presented a completely functioning dual system. This system has since become the basis of digital technology.

Another example of Leibniz's ingenuity was his improvement of the technology of door locks. He noticed that traditional locks had many weaknesses that could be exploited by burglars, and he set out to design a better lock.

Leibniz's plans for a submarine were also noteworthy, as they demonstrate how genius can develop ingenious concepts that may not find practical use. His design was not successful in his time because it was not practical for the technology of the era. The materials available at the time were not strong enough to withstand the pressure of the deep sea, and the propulsion system was not powerful enough to move the vessel quickly or effectively.

But this should not encourage you, because despite his lesser importance in the history of the development of the submarine, Leibniz's ideas were still innovative and unique. His plans demonstrated a visionary approach to solving problems and a willingness to think creatively about new solutions. Leibniz's accomplishments were not limited to mathematics and engineering.

He also developed a device for measuring wind speed, advised doctors to take fever regularly, and established a widows and orphans fund. These initiatives show that challenges can take on many forms and can lead to a range of accomplishments.

The most famous and successful geniuses have created more than one *output*. It is interesting to note that even after their great successes – after which one would expect them to retire - geniuses continued to write books, give speeches and lectures, and produce other *products*. Completing small projects on a daily basis can help you develop a habit of closure and ultimately lead to larger accomplishments in the long run. While larger projects take time, it is important to make progress every day – by doing so, you can establish a sense of momentum and cultivate a mindset of productivity.

> "The best way to predict the future is to invent it."
> – Alan Kay

In the course of learning, something has always emerged that we can grasp today. Usually this was a *book*, a *scientific paper*, or even an exchange of *letters* (correspondence) between the universal genius and other scholars. This is how we, today, can have an idea and see what these people knew – or at least what they knew after writing it into the *medium*. It is thus conclusive for us to understand *how* the person studied a particular concept or thing. Geniuses have used different means to present their ideas, such as books, newspapers, lectures, etc. Coming back to Leibniz, you can see this variety in his *output*:

- Concept of a means of transportation in the form of a *drawing*
- Construction of a calculating machine in the form of a *model*
- *Letters*, *lectures*, *books*, and so on.

Geniuses wrote various works, including *essays, articles, letters, reviews, commentaries,* or published their own – political or scientific – *journals.* Here is a list of types of what a format or medium could be: *Articles, almanacs, blueprints, letters, biographies, diaries, dictionaries, encyclopedias, essays, handbooks, manuals, journals, seminars, lectures, discussions, memoirs,* or *portraits.* A finished *product* could therefore be a *scientific article* (format) in a *journal* (medium) on the subject of *existential philosophy* (concept). I usually prefer to choose a *book* as a medium for my challenges, but you can choose whatever medium you like.

It is absolutely acceptable to choose many different mediums or formats and write or publish them too. You don't have to try to change the world right away, because that is not particularly necessary at the beginning. It is enough to expand one's own level of knowledge. Focus on creating more *output* and you will learn more than you ever thought you would.

You might also find critics of your work as well. Within philosophy, various schools of thought emerged, each presenting theories under a common framework: Hegelians found themselves divided into left and right factions, while proponents and critics of Kant's philosophy existed side by side. It is worth mentioning these examples, as even philosophers like Hegel and Kant held distinct positions on philosophical theories, supporting some while opposing others. This serves as a reminder that geniuses, too, engage with and contribute to the existing body of knowledge and opinions.

> „I think that productive work is one of the true goods of life; when you work productively, you manufacture more than money, you also manufacture a sense of self-esteem for yourself."[100]

You now understand the benefit of such a challenge and why studying complex concepts is important. What follows is a challenge that I had to deal with quite some time ago, mainly the challenge of not being imprisoned by my own mind and learning something new out of the ordinary – something that I would usually not deal with. You will see what is interesting in this particular case of a challenge as it was a) not set by myself and b) was not particularly difficult because the complexity was challenging to grasp, but rather because of the misconceptions I had about the topic.

Interlude:

---

[100] Maltz, 1978

Critique & Bias

The process of engaging with a particular theoretical framework is a complex and multifaceted one, influenced by a variety of factors, including but not limited to, the prevailing attitudes towards it. A major factor influencing whether or not we engage with a theory is the general opinion about that theory. For example, the general negative opinion towards Marxism is predominantly unfavorable, largely due to the perceived failures of communism and skepticism about its practical applicability. This aversion to Marxist theory can impede our efforts to delve into its complexities as we are already biased towards it. The intricate nature of these theories may also prove challenging to comprehend, leading us to avoid further engagement and instead rely on the assumption that others have already studied and understood them thoroughly. As such, the inclination to take the time and delve deeper into its complexities may be inhibited. Our assumptions and attitudes towards certain intellectuals are also influenced by various *critiques* and reviews. One such critique – a book that does not write particularly positively about various French intellectuals, philosophers[101], psychoanalysts and their works, such as Jacques Derrida, Michel Serres, Julia Kristeva, Bruno Paul Louis Latour, Jean Baudrillard, Gilles Deleuze, Felix Guattari, and Paul Virilio – is *Fashionable Nonsense*. This book was written out of motivation to further publicize the so called *Sokal Affair*, in

---

[101] The term *postmodern philosopher* and *Marxist philosopher* are often used interchangeably. I want to point out that, while I was talking about Marxism before, many of the philosophers criticized by the authors mentioned are more considered postmodernists rather than Marxists.

which a nonsense-article[102] written by Alan Sokal himself was accepted by a non peer-reviewed journal, and to brand as such the nonsense that these intellectuals would spout.[103] However, all that has happened as a result of this is, that this article and paper are cited over and over again by many people – mostly on internet forums – without having really looked into the criticized theories themselves. People make it easy for themselves and refer to titles like these and try to deny these intellectuals their status and declare their theories as nonsense.

> "This is all rather sad, don't you think? For poor Sokal, to begin with. His name remains linked to a hoax – "the Sokal hoax", as they say in the United States – and not to scientific work. Sad too because the change of serious reflection seems to have been ruined, at least in a broad public forum that deserves better." – Jacques Derrida[104]

The endeavor of engaging with the theoretical contributions of intellectuals necessitates a critical approach, not merely captivation with their writings. My own foray into the works of Slavoj Žižek, for example, began with my desire to scrutinize his theories, prompted by my interest in Sir Roger Scruton's book *Firebrands* – as I found out later a rather shallow critique on intellectuals and individuals such as Eric Hobsbawm, Edward P. Thompson, John Kenneth Galbraith, Ronald Dworkin, Ronald D. Laing, Jurgen Habermas, György Lukács, Jean-Paul Sartre, Ralph Milliband, Michel Foucault, Louis Althusser, Jacques Lacan, Gilles Deleuze, Alain Badiou, and Slavoj Žižek.

---

[102] Transgressing the Boundaries: Towards a Transformative Hermeneutics of Quantum Gravity

[103] Sokal & Bricmont, 1999

[104] Wolters, 2013

Even before forming an opinion on Žižek's work, I was motivated to read Scruton's book and engage with the former's ideas through a comparative analysis. A friend's recommendation that I explore Žižek's theories further, emphasizing the philosopher's skillful presentation of Marxist theory in an entertaining manner, further reinforced my resolve to approach these works critically. I would occupy myself intensively in order to be able to form my own opinion on them. Consequently, I invested a considerable amount of time and effort into studying Žižek's theories, attending Žižek's lecture in Vienna, watching YouTube videos, reading *The Sublime Object of Ideology*, and analyzing his ideas extensively. I was enthusiastic and continued to study his theories for years to come. I also read more from the book *Firebrands* later on, but my experience was somewhat disappointing, as I found it to be superficial and of limited practical use.

To write a really good and intellectually noteworthy critique takes some effort. To declare theories as nonsense is easy but not a valuable critique after all. However, this path is very often taken for the sake of simplicity, and the reader, already overwhelmed by the complexity of the works of various philosophers anyway, carelessly agrees with the critique without having really examined their theories more closely. You can see that this does not make very much sense — we should take the time to deal with complex theory, seriously engage with their works, and *go beyond simplicity*. As scholars, it is incumbent upon us to approach the task of knowledge acquisition with a critical eye and an unwavering commitment to the pursuit of truth. In the pursuit of understanding complex subjects, *critiques* play a significant role in shaping scholarly discourse and directing the attention of readers to particular works, but we must be cautious not to become blinded by the opinions of others, no matter how esteemed or influential they may be, as these evaluations may not always be accurate or legitimate. When it comes to Slavoj Žižek, I have my own findings and opinion that I have withheld for too long already. It may be beyond the scope of this book, but I do not want to withhold this information from you, the reader, as I think it can be useful should you ever have to deal with his books – or any other intellectual or philosopher that you want to understand. These are personal opinions and not rooted in any particular scientific method:

- Do not treat each and every book of them separately, but rather in connection with every other book they have written, i.e. treat the *Œuvre* as a whole – in a *system*[105].
- Deal with them as a writer that combines several theories into a new perspective that is never truly communicated as one coherent philosophy, but is actualized within, by reading and understanding the whole *system*.
- Understand the meaning behind such combinations and why they are important rather than knowing what each and every concept being used means in particular.
- Read the introductory books of them first, then read the most difficult one. With Žižek this could be *The Sublime Object of Reality* followed by *Less Than*

---

[105] More on systems in Part 3 of this book.

Nothing. With Marx this could be *A Contribution to the Critique of Political Economy* (*Zur Kritik der politischen Ökonomie*) followed by *Capital* (*Das Kapital*).

- Reading critiques of their work is never enough. Always deal with the subject at hand. Books that try to simplify take away the evident research that is so necessary for the serious scholar of their works. People who read such simplifications are going the easy way and will not be able to grasp the full complexity of their works.

This is a great introduction to the next chapter where we will deal with the knowledge acquisition and learning of complex concepts, as well as the definitions of what complex concepts are and what an understanding of them entails.

Introduction:
Studying Complex Concepts

What are these complex concepts I am constantly talking about and how do they relate to this book? What kind of science should we focus on? What is the all-relevant always necessary science that can find its use in every aspect of time and space? Since science is a broad area of study that encompasses a wide range of subjects, from biology and physics to chemistry and geology, it is hard to say *the* most relevant science.

> "such terms as "physics" or „biology" are fairly meaningless labels, and experts in particle physics would find it very hard to understand even the first page of a new research paper in viral immunology. Obviously, this atomization of knowledge has not made any public decision-making easier."[106]

Some examples of complex concepts to focus on include *quantum mechanics, abstract algebra, machine learning,* and *space engineering,* just to name a few. If we would only knew what to focus on, we could just grab a book and start learning. If everything is equally important, then it is equally unimportant. Why spend a lot of time learning something if you can't see the immediate meaning behind it? Prior to

---

[106] Smil, 2022

discussing the specific details of complex topics and their importance, I would like to share my personal motivation for studying them.

Lots of books have dealt with studying and time management skills in a high school or college environment, like *How to Become a Straight A Student* and *What Smart Students Know*. They focus entirely on making it as a student in the respective area, give a good overview on how to manage your time and how to study for exams. When it comes to solving specific problems, then *How to Solve It* is a suitable book for that matter and when it comes to solving mathematical problems, then *Mathematical Marvels: Adventures in Problem Solving* and *Solving Mathematical Problems* are also worth mentioning. Some books focus on the totality of thinking specifically for an area of study or work, like *Thinking Like a Mathematician* and *How to Think Like a Programmer*. A lot of value can be taken from these and many other books, such as the countless books written on the area of *Speed Reading*. Although many books center around learning and studying within an academic context, it is equally important to engage with complex topics during leisure time[107]. Even if one does not perceive an immediate necessity to do so, the benefits of expanding one's knowledge base extend beyond the bounds of formal education. Unfortunately, most individuals fail to recognize the significance of pursuing topics beyond their field of expertise. Those who do prioritize intellectual pursuits are often affiliated with institutions of higher learning. By engaging with complex ideas outside of one's usual sphere of knowledge, an individual can enhance

---

[107] Such as many geniuses did as we have previously seen.

their ability to think critically, sharpen their analytical skills, and improve their capacity for abstract reasoning. Moreover, the act of intellectual exploration is an excellent exercise in self-discipline and a crucial trait for success in academia and beyond. Yet, many people fail to do so because they believe that such endeavors are not pertinent to their daily lives. In this part of the book, the focus lies specifically on the understanding of complex *concepts* – their acquisition, their relation to systems, and why it is important to create your own concepts in systems if you want to improve your learning endeavor – rather than general learning techniques.

This is the chapter that tells you exactly what you need to do, what science you should focus on, what concepts to learn, and presumably how to learn them too. But is it really that simple? While there are certainly specific skills and concepts that are important to master in any given area of study, the process of becoming proficient in a particular field is often much more complex than a simple step-by-step guide can capture.

It is important to be aware of what skills and knowledge are relevant to interests and then focus on those. It is also important to remain open to new learning and seek out new challenges on a regular basis. We cannot always predict what we will need in the future and by building a broad base of knowledge and skills, we are better prepared for unforeseen events and new challenges. In order to learn new concepts, we must first discard old ones. This does not mean that we have to give them up when we have realized that they are actually the better choice – but not because we are used to choosing that way. It is difficult to learn to understand the difference.

Some people just like to learn new things because they are curious and want to deepen their knowledge. But how does one ultimately get the curiosity to study a particular science? Many factors could be decisive for this. Experiences in everyday life, such as observing natural phenomena or solving problems, can also lead to an interest in certain sciences. The influence of friends and family can contribute as well. Perhaps one should look online for a little inspiration? But the Amazon bestseller list has changed very certlittle in recent years. The same productivity books are still on the bestseller list, just as they were back then. That's not surprising, because the bestseller list continues to market bestsellers: Popular books that push existing opinions and propagate the same views over and over again. If you follow the self-optimization gurus on Instagram or anywhere else, you'll see that their opinions haven't changed much either.

In the movies *Limitless* (2011) and *Lucy* (2014), the mental capacity and cognitive abilities of the main characters are enhanced with the help of a drug. Similar drugs can also be found in *Dune* (Spice), *Star Wars* (Glitterstim), as well as other books and movies. The success and interest in such films suggests that the need to improve cognitive abilities is there, but that people are always looking for shortcuts to avoid the effort of learning. As already mentioned, there are always people who take the difficult route and choose to learn something that requires more effort.

Where does the interest in a topic come from? I have also spent a long time getting to the bottom of this question. My goal is not to present you with a few interesting learning methods, but to awaken in you your personal motivation to learn even more. We learn at work, in our free time, through news, events of everyday life, and much more. But learning complex topics requires a specific focus and – most of all – the individual's choice to do so.

If you always learn only what you are currently interested in, you will not be able to motivate yourself for very long. Because even if something may seem very interesting and enjoyable at first, it can be difficult to motivate yourself for it in the long run. What if this joy is simply gone? It can happen that certain things are only enjoyable at the moment – but as soon as they become more difficult, you no longer want to deal with them. We want to keep that motivation for a longer period of time and move past such obstacles.

What is the subject that you can fully emerge in, that you spend hours on without even looking at the time? Some are interested in computer science, teach themselves new programming languages, are interested in the latest news on computing, or just want to learn more about the subject because it really excites them. If curiosity grips you and you also keep reading through articles about physics, chemistry, biology, then maybe these are the subjects that really fulfill you. In such situations we are in a flow and afterwards much more fulfilled than before. The pursuit of such a thing would then probably include participating in panel discussions, reading many books in that particular field, visiting museums, and much more. So it's not the activity that gives immediate pleasure, but the engagement with the thing that excites you. Maybe start your own project and write a book about it?

Few will read about system architectures (monolith vs. microservices), thermonuclear astrophysics (nuclear physics and astrophysics), phenomenology (philosophy), or any other topic if they don't need it directly for their job. How is someone whose main part of life is doing an activity over and over again supposed to get a taste for a new topic if they've never heard of it? What motivates individuals to subject themselves to such rigorous and complex study as Mathematics?

Despite being regarded as one of the most disliked subjects during one's academic years, there still exists a subset of individuals who choose to pursue it in their lifetime. Many individuals often question the relevance and practicality of studying mathematics due to its perceived difficulty and lack of direct real-world applications. But despite the seemingly overwhelming arguments against it, mathematics serves a significant purpose in modern society and plays a fundamental role in various fields, such as science, engineering, economics, and computer science. But the applications are not only limited to that. It has been shown that mathematics is useful if you want to pursue an academic career. At the university level, most academic programs necessitate the inclusion of mathematics as a core component of their curriculum. This is because the proficiency in mathematical abilities is deemed as a crucial gauge of students' potential in all academic pursuits, irrespective of the field of study.[108]

No book could ever convince or motivate you to like mathematics, but that is an essential factor to study mathematics. I can't convince you that you should study higher mathematics. The reasons are numerous but too abstract to grasp. In fact, any reason I would give you is only my personal motivation. This has nothing to do with you. I

---

[108] Tang et al., 2009

learn because it *excites* me. A fascination for a subject can at most be described, but never explained. Therefore, it is not possible for me to show you why you should occupy yourself with - and also study - mathematics. What is my motivation to study more complex concepts? Since I have a huge interest in engineering, studying mathematics was a must for me:

> "In engineering, mathematical knowledge is one of the most important tools for engineers."[109]

In addition, I enjoy complex topics. I enjoy learning something that I don't yet understand. It makes me humble and give me something to look forward to. I really like the idea that there is still so much out there that I do not know yet. Just read the following passage:

> "We construct a Kirby color in the setting of Khovanov homology: an ind-object of the annular Bar-Natan category that is equipped with a natural handle slide isomorphism. Using functoriality and cabling properties of Khovanov homology, we define a Kirby-colored Khovanov homology that is invariant under the handle slide Kirby move, up to isomorphism."[110]

For some, this text might be intimidating. I can certainly understand that. But people are too afraid to say that they don't understand something. But that is exactly my motivation to deal with a topic.

It might possible that I enjoy reading such texts because they are challenging and intellectually stimulating. They require a high level of understanding and expertise in a

---

[109] Asshaari et al., 2012
[110] Hogancamp et al., 2022

particular field – something which I do not have yet –, which makes them fascinating and thought-provoking for me. It might also be the language and terminology used in such texts which is unique and complex by itself. A new topic with an element of novelty and intrigue to the reading experience. Precision and clarity of language, as well as the highly complex, all of this – and much more – motivates me to build interest in that area.

If you are not open to new ideas, you will never get them. You have to cultivate an open mind and learn things you may not need today. You never know when creativity will strike, but it comes when many ideas are combined. Look for things you don't understand today because they might give you ideas for what you need to learn tomorrow. Become a seeker!

> "Approach your own personal voyage and projects like Michelangelo approached a block of marble, willing to learn and adjust as you go, and even to abandon a previous goal and change directions entirely should the need arise."[111]

The more you learn, the more you are able to learn because you now understand concepts you didn't even know about before. A university degree is much more than just an education to earn more money afterwards. It opens up many more doors for people to further their education in scientific fields. It's not just about programming or biology or physics, it's about the ability to now move into subjects you couldn't get to before.

By opening up a new subject area, I now have the opportunity to get more involved with it. Because of the basis

[111] Epstein, 2019

that I now have, it is easier for me to deal with it on a deeper level later on. Should I continue delve deeper into the topic or should I also learn in other areas of which I know little or nothing at all? We can't yet know whether the knowledge we don't know about will benefit us in the long run. Because often it is not the knowledge per se, but the acquisition – the process of acquiring – of said knowledge that can drive our intellectual growth. By dealing with it, we learn to deal with difficult situations, to solve problems, and not to run away from challenges we face.

The most important ability of man is to be able to constantly learn something new. This is the only way mankind has been able to develop. That's why there are bridges, electric light, computers, books, smartphones, and strawberry tiramisu – after all, we should not ignore the culinary delights that are available today. By learning new concepts, I don't mean simply consuming information. Because there is a significant difference between something that is easy to understand right away and something that requires a little more effort. You can read an online article filled with truisms and still not learn anything from it. You can be reinforced in your own view by such articles, but nothing more. You may learn *about* something new, but you haven't necessarily learned something new. After all, learning that Khovanov homology[112] exists, does not mean that you understand it.

It is not just about being able to recall information. That is a side effect of learning, but it does not show whether we have understood much. If I learn factual knowledge and thus

---

[112] If you are here to learn more about it, you have the necessary curiosity to teach it yourself by going online and researching the required materials.

now know that African languages have been divided by Joseph Greenberg into four groups (Afro-Asiatic, Niger-Congo, Nilo-Saharan, Khoisan[113]) it may not bring me any immediate benefit. I will probably also forget it fairly soon. However, it may benefit me in the further understanding of this thing, in the long run.

> "Memorization creates the repertoire of what we think about. Nobody can think in a vacuum of information."[114]

But the ultimate goal is to understand the concepts and their meaning – not just being able to list them. What is this knowledge that is important? Who defines it? How is someone supposed to know whether knowledge he knows nothing about is important? Is it really only about learning something that you can use later? Then it might be better to concentrate on what interests you. But how can you know what you are interested in if you don't know it?

These are very many questions that show that there is no legitimate reason *not* to learn something, because we just cannot know today whether the knowledge in this matter might not be useful to us in the future. I would simply say this: *Learn more!*

> "Since we cannot know what knowledge will be most needed in the future, it is senseless to try and teach it in advance. Instead we should try to turn out people who love learning so much and learn so well that they will be able to learn whatever needs to be learned."[115]

---

[113] Correct term would probably be the Khoe-Sān, combining the Khoekhoen and the Sān – indigenous peoples of Africa who speak click languages.

[114] Kwik, 2020

[115] Attributed to John Holt

We cannot know whether the knowledge of which we do not know yet will benefit us, because often it is not the knowledge per se, but the process of acquiring that knowledge. By dealing with it, we learn to deal with difficult situations, solve problems, and not run away from them. Don't focus so much on the hurdle, but on the fact that mastering it will help you in the future. By mastering something, you are not mastering the hurdle, but mastering yourself. This is how you become a master.

> "The more that you read, the more things you will know.
> The more that you learn, the more places you'll go."[116]

By opening a field of study, one then has the opportunity to deal with another more   complex one. How can one understand quantum physics if one has no idea of higher mathematics?

Through a basic understanding of a subject matter, it is easier to deal with it even more precisely and more deeply. Later on, one asks oneself: Should one delve even deeper or should one also learn in other areas of which one knows little or nothing at all? Both are legitimate and important:

> "As the division of intellectual labour has increased, even polymaths have become a kind of specialist. They are often known as generalists because general knowledge, or at least the knowledge of many disciplines, is their speciality. Their distinctive contribution to the history of knowledge is to see connections between fields that have been separated and to notice what specialists in a given discipline, the insiders, have failed to see. In this respect their role resembles that of

---

[116] Geisel, 1978

scholars who leave their native country, whether as exiles or expatriates, for a place with a different culture of knowledge."[117]

In the pursuit of knowledge, there is no authority directing one's course of study or dictating the material to be learned. This is something we also have to learn, since nobody is telling us what we should pursue to study. Such a pursuit is not necessarily driven by an external incentives, but rather by a deep-rooted desire to better understand the world and its many intricacies.

Yet, the lack of clear direction regarding what to do and what to learn can pose a challenge for many individuals, particularly those who are accustomed to being given explicit guidance. It requires a certain level of self-motivation and determination to chart one's own course of study and engage with complex topics. Only by exploring complex topics on their own, individuals can develop a deeper understanding of the subject matter and gain insights that might not be accessible through other means.[118]

But you first have to overcome this obstacle of not having a clear roadmap to truly gain all the benefits this journey holds for you. Just learning more and engaging with complex topics does the trick, as the act of encountering and grappling with complex topics can spark a sense of curiosity and interest, ultimately fueling a desire for further understanding and exploration. This process of intellectual growth and engagement may be facilitated by seeking out diverse and challenging sources of information, as well as by approaching new topics with an open and receptive mindset.

---

[117] Burke, 2021
[118] Ditta et al., 2020

Despite the availability of information and resources, many people remain disengaged from complex science and education. This raises the question of how individuals choose their areas of focus and interest, and whether the media they consume plays a role in shaping their perspectives. I often ask myself if it would be different if shows on TV would focus some more on complex science topics. News media, in particular, can have a significant impact on what people perceive as important, and what topics they choose to explore further. By providing a more accurate and comprehensive view of the world, news media could encourage individuals to become more engaged in complex science and education.

It is important to recognize that our perceptions are not objective or unbiased, but are rather shaped by our experiences and environment – and this includes what you watch on television or the books you read regularly. News media plays a crucial role in shaping public opinion, and can influence what issues and topics people consider to be important. But it is not just the news, it is also what you read online[119]. This means that what one individual considers to be interesting may not be the same as another individual's perspective. I just want you to realize that your perception of what is interesting or important to you is influenced accordingly.

Therefore, your perception of what you might consider important is somewhat biased. Confirmation bias is one such common bias in which individuals seek out information that confirms their existing beliefs, while ignoring information that contradicts their beliefs. This can lead to a narrow and

---

[119] Chang & Tsai, 2022

biased view of the world and stops people from learning something new or prevents them from considering alternative perspectives and viewpoints.

This means you will most probably not learn anything new about mathematics, neuroscience, cloud computing, machine learning, or aerospace engineering, if you do not actively seek out sources accordingly. So you have to evaluate what you constantly consume and what you think is interesting, because nobody will tell you what is. I am often thinking about how different my life would be if I was raised in a different household, surrounded by books on topics way out of my intellectual capabilities. While we cannot change the past, we can change how we perceive our life today – and therefore, change our future.

> "Research suggests that thinking about the future—a process known as *prospection*—can help us lead more generous and fulfilled lives."[120]

As we learn more, we understand and develop an ability to see technological trends, understand how they work together, can instantly and intuitively see what will work and what will not. This alone is a valuable skill to have to sustain successfully in this constantly changing environment. You can't just rely on what you read in the newspapers.

> "Quantum computing will revolutionize the world. If you trust the headlines. Which you shouldn't." – Sabine Hossenfelder[121]

---

[120] Berkeley, 2023
[121] https://www.youtube.com/watch?v=IhS6ecYZFdQ

One must develop a discerning eye to identify trends in science. It is not advisable to rely solely on media coverage as it may be biased or sensationalized. Instead, engaging in continued learning and staying up to date with the latest developments in the field will enable one to better understand and evaluate scientific trends. Some may be more significant than others, depending on their impact on the scientific community and society as a whole. One must develop an in-depth understanding of the field and be able to critically evaluate the significance and relevance of new discoveries and developments. By approaching trends in this manner, one can distinguish between genuine advancements and hyped-up media stories. A good way to get to know what sciences could be relevant in the future, is exploring science fiction media.

> "Over the last thirty years, tens of thousands of stories have explored all the conceivable, and most of the inconceivable, possibilities of the future; there are few things that can happen that have not been described somewhere, in books or magazines. A critical-the adjective is important-reading of science fiction is essential training for anyone wishing to look more than ten years ahead. The facts of the future can hardly be imagined ab initio by those who are unfamiliar with the fantasies of the past."[122]

While science fiction is usually considered a form of entertainment, it can also be used as a tool for educating and inspiring people about science and technology. With this perspective I want to explore the potential of science fiction as a medium for learning and understanding science, particularly in the context of emerging technologies.

[122] Clarke, 1962

Science fiction is a genre that imagines possible futures and the scientific and technological advancements that may shape them. Through speculative narratives and fictional worlds, science fiction invites readers to consider the consequences and implications of new technologies. This is particularly relevant in our present moment, where technologies such as artificial intelligence, biotechnology, and nanotechnology are rapidly transforming society and raising critical ethical and social questions.

One way that science fiction can enhance scientific learning is by providing characters with a scientific background. Science fiction narratives often involve futuristic technologies that are not yet a reality, but which are based on real scientific principles and the studies needed to understand these principles. Hence, reading and watching science fiction can provide inspiration and motivation for exploring science, technology, engineering, scientific education and engineering careers further. It often features characters who are scientists or engineers, and who use their knowledge and skills to solve complex problems and create new technologies. These characters can serve as role models for individuals interested in science, and can help to break down stereotypes about who can be a scientist or engineer.

One of these characters that I personally find very inspiring is Tony Stark (*Iron Man*). To be like Tony Stark you would need to study *Aerospace Engineering, Nuclear Engineering, Mechanical Engineering, Electrical Engineering,* and *Computer Science.*[123] That already gives me a lot of

---

[123] Answer to the question **What kind of education should one follow to be like Tony Stark?** On Quora. Sumontro Sinha gave a wonderful and extensive answer, highlighting all the skills and knowledge necessary to achieve this endeavor.

possibilities for possible further education. However, sometimes it's not the heroes with more brains but the villains. – such as Doctor Doom, who "is considered one of the most brilliant minds and scientists on the planet Earth"[124]. He is also known as being "a sorcerer with skills in magic, matching the most powerful beings in the Universe"[125].

> "Any sufficiently advanced technology
> is indistinguishable from magic."
> – Arthur C. Clarke[126]

It is important to mention that science fiction is not a substitute for scientific education and should not be taken as an authoritative source of scientific information. However, it can be used as a great source of inspiration for whatever kind of technology you find fascinating and could give you just enough motivation for you to study this technology, i.e. focus on learning something complex and going *beyond simplicity*.

---

[124] Marvel Fandom - Victor von Doom (Earth-616)
[125] Ibid
[126] Clarke, 1962

Abstraction:
Concepts and Systems

———

But before we start to deal with complex concepts, I want to explain a little bit about *abstraction* and give a few examples and difficulties too. To understand the *universe*, you have to understand *concepts* and *systems*. The understanding of *concepts* and *systems* facilitates or even enables thinking in *concepts* and *systems*. This is extremely important because it makes it easier to grasp knowledge of the world. In addition, such a *way of thinking* helps us to analyze information and makes it easier for us to think abstractly. It is therefore the case that the knowledge of *concepts* and *systems* not only brings us understanding of what these terms mean – we can also learn to *think* in terms of them. You later might conclude that all words that have meaning describe a thing that *happens*[127] in the real world. Anything that is not in your immediate vicinity must be explained in terms of *concepts*, and concepts are quite *abstract*. So to understand the universe, you have to understand *abstract* thinking, or rather, you have to be able to *think abstractly* – that is in terms of *concepts* and *systems*.

What many if not all *complex concepts* have in common is that they are quite *abstract*. In fact, every time we think, we are

---

[127] Why do I say "happens?" Because things are not, they happen. When you say, "This is an apple," the apple does not remain, but it is *happening* right here at this moment in time and space. In a hundred years, it won't be, but that is another issue.

thinking in *abstractions*.[128] Most of what we know is *abstract*[129]. So to understand reality, you have to be able to think abstractly. Everything we know about the reality is an abstract reality – a *concept*[130]. In fact, most of what we know is *intangible*. Everything we talk about becomes abstract when we talk about it. It becomes a *representation* of reality, but it is not *reality* itself. Even if you can see things, e.g. inspecting cells under a microscope, they become an *abstract* concept when you talk about them and their behavior. By the time you talk about them, they may or may not behave that way – sometimes we do not know.[131] When you describe what you see, you immediately place the *object*, the *thing*, the *situation*, into a representational model of how you would *describe* that thing[132]. That doesn't mean that your description is hundred percent consistent with reality[133], even if it is happening at that moment. You would always say, "I have a banana at home," but you would never say, "I have *the* banana at home," because that would mean that there is a particular banana that has certain properties and is unique in the *banana*-category.

"They speak of existence of *external* objects, which can be more precisely defined as *actual*, absolutely *singular*, *wholly personal*, *individual* things, each of them absolutely unlike anything else; this existence, they say, has absolute certainty and truth. They *mean* 'this' bit of paper on which I am writing – or rather have written – 'this'; but

---

[128] Thinking that we think is an abstraction of our thinking – which is itself a process of abstraction. Thinking that we think we think is an abstraction of the abstraction of our thinking, and so on.

[129] It is a *concept*.

[130] A concept is often used to help people understand and organize the world around them by providing a way to categorize and make sense of complex information, but not every concept has a representation in reality.

[131] This happens quite often in quantum physics.

[132] We understand that this description is limited by the language we use.

[133] It probably is not as we are all biased too.

what they mean is not what they say. If they actually wanted to *say* 'this' bit of paper which they mean, if they wanted to *say* it, then this is impossible, because the sensuous This that is meant *cannot be reached* by language, which belongs to consciousness, i.e. to that which is inherently universal."[134]

Of course, every banana is a clone, but that is another story[135]. When you buy things, you are not buying the thing itself, but the *concept* that is applied to the thing[136]. There can be many concepts applied to a thing: A *game* (concept) that can be played by *four* (concept) *players* (concept) is different from a *game* (concept) that can be played by only *two* (concept) *players* (concept). Yet, we consider both as one *game* (concept). Why are they both considered the same concept (a game) when they are so different from each other?[137] If we start to analyze each word and consider each a concept, it becomes even more difficult. What is the *first* thing that comes to mind when you hear the word *game*? That is your idea of what a game is – probably it has an object in reality but this does not have to be the case. You see, when you hear the word "game" and you think of "chess", then it is true that there is a thing outside of your imagination, i.e. in reality, that is considered a game or that you take as a direct representation of the word, the concept, *game*. The difficulty arises when you start with more abstract concepts like *speed* or even *speed of light*[138] – which can be seen as a combination of

---

[134] Hegel, 1977 – page 66

[135] Arias et al., 2004

[136] You are also buying ideology.

[137] This can be especially difficult to grasp for people with autism.

[138] I chose these examples because they are quite common and you might have heard of them even if you do not have studied physics at university. There are far more difficult concepts in mathematics which we will cover later on.

two concepts. These are things that cannot be grasped immediately because they are not *objects* in reality, but rather *phenomena* that take place in reality. When you think about *four* (concept) then you might think of *four* dots or *four* apples or just the number 4 – but *four* does not exist outside your imagination. It is an abstraction of reality. This can be *four* apples, bananas, trains, but as soon as you declare a differentiation within a certain category, i.e. you consider only apples with certain properties as apples of a category, you might have less than *four* of such apples, even if you had *four* apples previously. As a child, you were given an apple and were told, "This is an apple." This could mean that an *apple* is everything you can touch, feel, comprehend – which is everything that exists in your immediate reality –, but that is not the case. If we draw a maple tree, walnut tree, birch, spruce, and so on, it is always about the underlying *idea* behind it, namely that of a *tree*. However, the *idea* of such a *tree* can only be represented conceptually, that is, in the form of a simplified drawing, sketch, and so on. We will not find the concept "tree" in reality, because no generalization of the idea *tree* exists in nature. We use *systems* to describe these concepts – philosophical systems and metaphysics for abstractions and natural sciences considering natural phenomena –, as well as their relation to each other, as it is the case in certain relations which are called *categories*. Here, Hegel's system is also quite valuable. Later, we shall examine what exactly *systems* are and how they *function*.

We have seen that there is a difference between different *types* of *concepts*. For these types we must distinguish between a concept as an *abstract idea* of a certain thing of the reality (*immediate*) or a *new idea* which does not occur in our reality (*abstraction*). Therefore, a concept can be called either *abstract* or *immediate* – but this does not make the concept more abstract or more immediate by itself, as every concept is always abstract, i.e. an abstraction of something. It is much more about what a concept refers to, what it is related to, that is, what the *object* that this concept refers to *is*. If the object is a thing that is happening in reality, it is *immediate*[139], otherwise it is an *abstraction*. Thinking is itself a *process* of *abstraction*. An *abstract* concept is a concept that represents an *idea* or *thought* that is not tied to a specific physical object or sensory *sensation*[140]. This makes them a bit more complex and usually a bit more difficult to understand than so-called *immediate* concepts. They are often used when describing complex ideas or to help us understand more abstract or intangible aspects of our world. So it can be said that they are itself *complex* and describe *complex* ideas – which is exactly why it is important to understand them and why I mentioned them first. Mathematics is mostly concerned with *abstract* concepts, or *abstracted versions* of concepts that could have been attributed to other more *immediate* objects. Physics, on the other hand, would rather describe the more *immediate* concepts - concepts that relate to actual natural phenomena in our reality. We will return to both of these topics later in the book.

---

[139] An *immediate* concept is a concept that represents something directly related to a particular physical *object*. So this would mean a chair, a tree, or a person, but also sensory experiences such as the taste of a particular food or the feel of a particular texture.

[140] Examples of abstract concepts are justice, love, and intelligence. These concepts are not directly linked to any particular physical object or experience, but rather are ideas that we can think about and discuss.

> "Perhaps fresh philosophical insight could demonstrate that supposedly opposing points of view could be reconciled, that previous disagreements were based on misunderstandings, that the truths implacably defended by warring camps could in fact be harmonised with one another. Perhaps the language central to the new, mechanical philosophy - the language of mathematics - could even provide the prototype of a new philosophical language capable of resolving disputes of all kinds conclusively."[141]

A concept is *too complex* when it becomes overwhelming or unmanageable and when you are unable to understand it or use it effectively. This can be the case when a concept is too abstract or theoretical, when it contains too many interconnected parts or ideas, or when there are no clear definitions or examples. In these cases, it can be difficult to gain a complete or accurate understanding of the concept, and you may feel frustrated or overwhelmed when trying to learn or apply it. Generally, a concept is too complex to understand if it is beyond your current knowledge, skills, or abilities and you cannot make progress in learning or applying the concept. That does not mean you would not be able to understand these concepts if you would take the time to learn them.

Complex concepts are also put together in a *system*, that is to say, a complex concept is by itself a *system* of different *concepts*. Geniuses have always used various *systems* to recategorize the knowledge of this world and their own in a *system*. A *system* refers to a particular arrangement, categorization, and so on, of objects, *ideas*, or – more generally speaking – *concepts*. It allows us to organize and understand complex phenomena and how they relate to each other. Systems are used in many different fields, from biology and

141 Antognazza, 2011

engineering to economics and are essential to making sense of the complex world around us. By defining and analyzing a system, we can better understand how the parts of the system *interact* with each other and how they *contribute* to the overall *function* of the *system*. *Systems* provide a *framework* for understanding how the individual parts of a system interact with each other. This allows us to identify problems or inefficiencies in the system and develop solutions to improve its performance. In addition, systems can help us better predict and control the behavior of *complex systems*[142], which can be useful in many different contexts. So understanding systems or even complex systems help us far more than just to the immediate understanding of the system we want to relate to. We refer to this as *Systemverständnis*[143]. In order to be able to understand systems in every possible form, we need to define some concepts first.

> "A system is an organized group of related objects or components; models can be used for understanding and predicting the behavior of systems."[144]

A *model* is a specific representation of a concept and is used to understand or predict some aspect of the concept. For example, *evolution* is a broad idea that explains how species change over time. An evolutionary model, such as the theory of natural selection, is a specific explanation of how evolution occurs. A model is therefore called a *representation of a system* or process and is used to understand, predict, or explain

---

[142] There can be complex concepts and there can be complex systems.

[143] Understanding of systems is the ability to see the immediate connection between other concepts and theories in a *system*.

[144] NSTA, 2014

certain phenomena – which can be *immediate* or *abstract*. It may be a mathematical equation, or a computer simulation. A model is often used to test hypotheses or to make predictions about the *system* it represents.

Knowledge is ordered in a special form, namely in that of a personal system. This can either be inherited from different systems, be a completely new system, or be an existing improved system. Knowledge refers to the facts, information, and skills that a person has acquired through training, experience, or study. It is the result of learning and acquiring information about the world. We can use our knowledge in many ways, for example, we can use our knowledge to solve problems, make decisions, or learn new things, that is, *complex* concepts. The more we know the easier it is to learn something new. We can also use our knowledge to communicate with others by sharing our ideas and expertise. In addition, we can use our knowledge to improve our lives and the lives of others by applying our knowledge to help others or by using our knowledge to create new products or services that benefit society. The way we organize and use our knowledge ultimately depends on our individual intentions, interests, and abilities.

Understanding, on the other hand, refers to the ability to think about, interpret, and make sense of information and ideas. This includes the ability to recognize connections and relationships between different pieces of information and to use that information to make judgments or decisions about the information at hand. Clearly, we need both *knowledge* and *understanding*. In fact, I would argue that there is no path to true knowledge without understanding. To know facts without understanding them is merely to repeat the facts without understanding the concepts behind them. To be able to explain requires understanding. Trying to explain something, however, can help you to understand it better yourself.

It is possible to *combine* systems to create a *complex* system. This can be useful in cases where multiple systems are interrelated or overlapping and where a more integrated approach is needed to understand or address a particular problem or challenge. For example, in ecology, multiple systems can be used to understand interactions among species, how ecosystems function, and the impacts of human activities. By combining these systems, it is possible to gain a more holistic understanding of the complex web of relationships and processes that make up the natural world. Combining systems can be a challenging and complex process that requires careful planning, coordination, and collaboration. There are many *complex systems* in science, including both natural and artificial systems. Understanding these systems requires a combination of analytical skills, knowledge of the specific system and its components, and the ability to think abstractly and view the system as a whole, since systems can involve multiple levels of abstraction, with different parts of the system operating at different scales or levels of detail. Such systems often operate in dynamic environments, meaning that they are constantly changing and adapting to new inputs and conditions. Some examples of *complex systems* in science are: The human body, the universe, computer networks, neural networks, societies (consisting of individuals and organizations), transportation systems, ecosystems (comprising all living organisms in a given area), cells (basic units of life and themselves complex systems), and so on. By understanding how these systems are structured, we are able to understand what the structure of systems is – this is what I call *Systemverständnis*. One system alone may not be enough to do this, but understanding many systems helps us draw analogies and later develop our own systems.

Complex *abstract* systems are collections of interrelated concepts, principles, and ideas. These systems often involve *multiple levels of abstraction* and may involve complex relationships and interactions among the various components of the system. Examples of complex abstract systems include theoretical models in physics or mathematics, philosophical theories, or legal systems. Because of their complexity, these systems can be difficult to understand and work with and may require specialized knowledge, tools, or techniques to study or apply. Despite their complexity, however, complex abstract systems can provide valuable insights and perspectives on a variety of phenomena and can be an important source of knowledge and understanding. Knowledge and understanding are closely linked, as understanding is often based on knowledge about a particular subject. However, *knowledge* alone is not enough to truly *understand*, as it also requires the ability to think critically and creatively. There is not "one" system for our lives or life itself, because life is a complex and dynamic phenomenon that can take many different forms. Different people, cultures, and disciplines may have their own systems or frameworks for understanding and organizing life, but there is no single, universal system that applies to all aspects of life. In biology, for example, life is usually organized according to the principles of evolution and the hierarchy of living organisms, from cells to ecosystems. In sociology, life can be understood in terms of social structures, relationships, and interactions. In philosophy, life can be studied in terms of consciousness, meaning, and purpose.

Each new *system* completes our understanding of the universe and of life itself. On the other hand, already established systems might blur our understanding and knowledge of the world depending on where these systems come from. How certain concepts and systems come into being would be a question for sociology, but this book is not concerned with that. According to Michel Foucault, every knowledge also presupposes a power relationship, and thus what we call knowledge today or tomorrow is dependent on the omnipresent power structure. But if knowledge is constituted by power and, in contrast, power is also influenced by knowledge, is not the concept of power - our current understanding of what power is, can, does – also dependent on our knowledge of it? According to this, we would not be able to take adequate measures at all to escape from this misery. It seems to me that this topic needs another book, in which I again deal with the prevailing ideological and questionably constituted concepts. Let's look at some other difficulties regarding *concepts* and *systems* when it comes to *understanding* them.

When the same terms are used for different concepts, it becomes confusing. For example, "*boilerplate*" refers to a recurring text module, standard wording or clause in contracts and documents in general. However, this word is also used in programming (*boilerplate code*) for code segments that repeat in several places with little or no variation. Here is the general concept "*repeated text block*" and it is applied to code and this is relatively easy to understand. The *word*[145] and the *concept*[146] are very similar in both cases. However, there are examples in other sciences of the same word being used for two different concepts.

In mathematics, a *field* is an algebraic structure, as is a group and a ring. But these last two terms (Group and Ring) have quite little to do with our usual understanding of a *group* and a *ring*. What we would usually understand as a *group* (a collection of persons, objects, and so on) would rather be equated with a *set* in mathematics. So the concept would be the collection of *elements* and the corresponding term in mathematics would be a *set*. With an *algebraic structure* one means a *set* with *operations* on this *set*. It becomes difficult when you try to find meaning in all of that: An algebraic group has nothing to do with a group which is a collection of people. However, a collection of single elements is not called a group in mathematics, but a *set*. A group in mathematics is the combination of a set and arithmetic operations. A *group* in mathematics is an abstract structure – an algebraic one – just as a ring or a field is such an algebraic structure. It is easier to imagine how these names came about: Because one found such structures again and again, one distinguished them uniquely from the others. So it was the *thing*, which is designated, before the *word*, which it designates. But mostly we learn the words first and then the explanation behind what it is supposed to designate. This complicates understanding immensely, in my opinion.

[145] the term denoting the *concept*

[146] in this case, a repeated text block

Some names of concepts make sense, others do not. Some have been named that way "in honor of", others refer to something else. Therefore, one should understand what other concepts there are for which there are either designations, or not. The question is why there is a designation for those and not for others and whether new designations should be given. The clear answer is: yes, new designations should be given if they make sense. What other designations there are will be learned first - before one learns what concepts there actually are. But there are also things without designations. These are probably not yet relevant enough, otherwise they would have a designation. Nevertheless, we use the formal language of mathematics, that is *logic*, for this, using mathematical *symbols*.

When I speak of *concepts* I mean the ideas behind them, not the *terms* or the *label* that describe those ideas. So when we learn *concepts*, we always refer to the original *idea*, not the actual *term* that denotes it. Let us look at the following two statements:

- To become a *successful* entrepreneur you need *willpower* and *creativity*.
- If the *cardinality* of the *union* of two *sets* is *equal* to the *sum* of the *cardinality* of these two *sets*, then the *intersection* of the two *sets* is *zero* – they are *disjoint*.

Truly, these two statements are very different in their informational content. We will not devote ourselves so much here to the truthfulness of both statements – the second statement is true in any case, the first statement is known as a truism – but to their *complexity*.

The first statement, as found in many self-help guidebooks, is undisputed in terms of its truth content. Probably no one will argue against it if asked for their opinion on this statement. Of course, other qualities would probably be necessary to become a successful entrepreneur, such as patience, resilience, resourcefulness, motivation, implementation power, and so on. However, the more we list such traits, the more likely it is that the inclined reader would oppose them. So the obvious thing to do is to limit ourselves to a few such qualities. What happens as a result is that the reader will definitely agree – which is what a writer in the self-help niche would want – although reader would not write a positive review, at least not a negative one.

You will be able to understand and comprehend the second statement if you have attended a course or module on basic mathematics. Let us take a closer look at the individual terms, explain them, so that you can understand this statement in any case and thereby also find its truthfulness: A *set* M = { 1, 2, 3 } and a *set* N = { 4, 5, 6 } have no elements in common. The cardinality of a set is expressed by the number of elements in just that set. Therefore, both sets have cardinality 3. The union of both sets would be { 1, 2, 3, 4, 5, 6 }. Therefore, the cardinality of the union of these two sets is equal to the sum of the individual cardinalities of both sets and since they have no equal elements their intersection is empty and their cardinality zero. The situation would be different for the sets A = { 1, 2, 3 } and B = { 3, 4, 5 }. Here each set would also have the cardinality 3 but the union would be { 1, 2, 3, 4, 5 } and this would have the cardinality 5. The intersection in this case would be { 3 } with the cardinality 1.

Coming back to the statements before, let us analyze their *difference* to each other. The difference between the first and the second statement lies not only in the complexity of the information, but also in the logical conclusion of its content. In the first statement, there was no logical conclusion but only one that made sense on the face of it – a successful entrepreneur who possesses just the following characteristics is not necessarily successful because he possesses these characteristics. In addition, not every person who possesses these qualities has to be a successful entrepreneur. This is also called a logical fallacy (*non-sequitur*), especially since the eclectic theory of trait and the Great Man Theory are subject to countless criticisms, neglect many other factors – as a simple trait does not guarantee success. Unfortunately, self-help books, guidebooks and the like, are full of statements of the first type. The problem that arises is this: You won't learn anything new from self-help books. In the end, they're not the things that give us the big insights, but they are marketed to us as if they would do. And that is a problem... It seems that if we keep reading these articles and books, we are spending our time wisely. Instead, maybe we should ask ourselves if we should really continue to do so. We think we are dealing with great wisdom, but in the end we are not learning anything new. When something like that is easy for us to understand, this is probably not because we are highly intelligent, but because we have read something that we already know. New things are often very difficult to understand and grasp. Therefore, it takes at least a little effort. Self-help books and articles often contain advice that is already known and lacks substance. People often skim through several articles without retaining anything useful. To really learn something new and valuable, you need to focus on *complex* topics and *theories* that take some effort to understand.

The problem with the self-help genre is that it presents obvious truths as great wisdom and fools people into thinking they are gaining knowledge, when in fact they are not learning anything new. To truly expand one's knowledge and understanding, one must go beyond such guidebooks and explore unfamiliar topics. Reading theories that take time to digest is more valuable than consuming endless amounts of easily digestible content. I can spend hours going from one interesting article to another without ever getting tired of it. It's just a huge pile of interesting things that aren't connected. It's like I'm visiting a lot of websites on topics that interest me, but in the end, I just spent a lot of time consuming information without learning anything really relevant. I might read an article about "10 Tips to Improve Your Writing Skills" and agree with everything the author says, but I haven't learned anything new. That is because I was already familiar with these things. Most of what you read in self-help books will already be familiar to you too! Do you want to continue like this or do you want to take your knowledge to the next level and actually learn something? If you are like most readers on the Internet, you go from one article to the next in one sitting. You read several articles, don't take any notes, and after an exhausting hour, you remember .... *nothing*. If you want to learn something new, you have to deal with things that are not obviously easy to grasp. So you have to be able to think *abstractly*.

Thinking:
Concepts and Systems

———

In the previous chapter we have dealt with the complexity of what *concepts* and *systems* are and how they are related to each other. In this chapter we are dealing with the *process* of abstraction, i.e. *thinking* in *concepts* and *systems*. First you need to learn how to learn new concepts, that is, understand how complex concepts connect basic concepts, the system of complex concepts, the underlying principles of concepts later used in complex concepts, and so on.

You cannot learn more difficult or complex concepts without understanding the basic concepts. Therefore, first you need to learn the basic concepts and assumptions that underlie the complex concept that you want to learn. Since I have a great enthusiasm for aerospace engineering, I will use rocket engines as an example here. There would be different types of such propulsion systems, such as:

- *Chemical rocket engines:* The most popular type of rocket propulsion systems are the chemical ones. To generate thrust, high-pressure gases produced by the combustion of propellant and oxidizer in a combustion chamber are ejected through a nozzle. Liquid oxygen and hydrogen are the most commonly used propellants in chemical rockets because they have relatively high specific impulse (a measure of a rocket's efficiency). Some rockets also use solid fuels, especially for their boosters.

- *Electric rocket engines:* These use electrical energy to accelerate the propellant to generate thrust. The propellant in these systems is usually a gas or plasma, and the particles are accelerated to high speeds by an electric or magnetic field. Electric propulsion systems are more efficient than chemical rockets and can produce much higher specific impulses, but they produce less thrust, so they are typically used for applications that require long, continuous thrust, such as satellites or space probes.

- *Nuclear rocket engines:* The working fluid in nuclear thermal rockets (NTR) is liquid hydrogen heated to a high temperature in a nuclear reactor and then expanded through a rocket nozzle to produce thrust. Because of the external nuclear heat source, this type of rocket engine has the potential to achieve a much higher effective ejection velocity and significantly increase payload capacity compared to chemical propellants.

Learning new concepts requires a combination of curiosity, determination, and persistence. If you are open to new ideas and willing to challenge your existing beliefs, you can expand your knowledge and understanding and be better equipped to navigate the world around you. The most effective method for learning new concepts ultimately depends on your personal learning preferences and style, as well as the particular topic you want to learn. Now, you would first learn to understand each of these propulsion systems before moving on to more complex concepts that require the understanding of such propulsion systems. Use Google, Wikipedia, or ChatGPT to find out everything you want. In addition to this, watch lectures or movies about it or talk to individuals who are knowledgeable about the concept to accomplish this. Finally, think about what you have now actually learned and how the concept fits into your overall understanding and expertise. This may mean making connections to other ideas or thinking about how the idea can be used in different contexts.

Then it is important to clearly define the new term and explain its meaning to ensure that it is understood by others. This means explaining exactly what the term encompasses and, more importantly, what it does not encompass. I have been somewhat vague in the past with my explanations of various concepts and as a result they have not carried through into my other books. It wasn't until I had a clear picture in my own mind of what I was actually trying to communicate with it that the application of the term was also clearer to me. In these cases, I used the terms output and, for example, products over and over again in context. This makes it easy for me today to explain what I probably wanted to say at the time. Because that's the difficulty: we try to put into words what is intangible in order to communicate it so that someone else or oneself, at any time in the distant future, could understand that term.

Today, however, I know that this is not possible without certain issues. Because concepts do not stand for themselves, without context, and especially not without context of the social circumstances in which they were created. This, of course, makes it incredibly difficult to write a *timeless* work. On top of that, technology, for example, is also constantly evolving and old technologies are being discarded. Since I already knew this, I rarely mentioned technologies in the past, but rather described in abstract terms what I actually wanted to say. Of course, this doesn't help the understanding of such texts at all and so today I can't fully understand what I meant in my books back then. But how can other readers understand if I don't know myself?

I have devoted several books to the concept of *output* and have mentioned it in almost all the books that relate to *becoming a genius*. Nevertheless, I have the feeling that much more could be written about it – especially what is involved in creating such an output, how to create one, what properties it has, and so on. I also keep using word order and so on to suggest that I meant these previous examples to be only exemplary and never conclusive. But with this happens another challenge: How large is the set of words that would still follow here and which would they be? Depending on the context, one could or would interpret other terms into it accordingly. As you can see, defining new concepts is not enough. You also have to make sure that they are understood. This may require breaking the complex concept down into its parts and explaining how those parts relate to each other – that is, understanding the *system*. It may also mean giving examples or illustrations of the concept as it applies in the real world to make it easier for others to understand.

Information rarely occurs in isolation. Our brain automatically looks for patterns to connect what is *new* with what is already *there*[147]. Using one's own senses or spinning a network of thoughts are natural methods that help. An important point in remembering information is the fact whether it was understood at all what is learned. Not enough, it is of course even more helpful to understand why it should be remembered in the first place, i.e. also by importance.

The *Massive Comprehensible Input*–method (MCI) is a natural learning method. This means that the brain connects the information by itself (*automatically*) in a way that makes sense to you. This makes information and words from foreign languages easy to remember. *Comprehensible input*, according to the linguist Stephen Krashen, refers to content that is understandable – Krashen originally used the term for foreign-language words whose meaning must be revealed to the learner, whether through a graphic, a translation, or other assistance. With MCI you strive to absorb as much learning material as possible, enabling a more organic formation of the thought network and better retention of knowledge

Whether or not this is the best learning method remains to be seen. The best learning method is of course a subjective matter, depending on the learning *style*, *goals* and *preferences* of the individual. Some people prefer to learn through hands-on experience, while others prefer to learn by reading or listening to lectures. Still others may prefer a combination of learning methods. Ultimately, the best learning method is the one that is effective for the individual and helps them retain and understand the material. It is also important to find a learning method that is engaging and fun, as this can help to motivate and encourage learning.

---

[147] It creates a new *system*.

Using a *clockwork* as an example, let's look at the different parts (*concepts*) and how you would learn them. If you were given all the parts of a watch, it would probably be extraordinarily difficult to put together a complete *movement* if you did not know the relationships between each and every part of the *movement*. So let us imagine that we are given a completed *movement*; we inspect it more closely by taking it further apart and see the *driver*, the *wheels*, the *escapement*, and the *controller*. We take each part aside, name it, and note how it is connected to the others. After we have taken the complete movement apart, we have the individual parts in front of us. Now it is up to us to put them back together in the correct order to construct a finished clock. This whole process of taking apart and understanding it is called *reverse engineering*.

> "Reverse engineering is the process of analyzing a subject system; first to identify the system's components and their interrelationships and second to create representations of the system in another form or at a higher level of abstraction."[148]

These terms have not been chosen by me personally. They have been defined by people some time ago already. The reason is simple: A common understanding about the composition of a clockwork enables us to describe it in an abstract form. By this I mean that we do not need to have the *movement* at hands in order to explain how it works. Instead of *this* does *that* and *that* does *this*, we have a clear language that allows us to share our knowledge with others – even without them ever seeing a *movement* mechanism in the real

---

[148] Jain et al., 2011

world. This is how teaching becomes possible. If we had to show a real object for everything we explain, we would have to bring earth, animals, and many other things into the classroom. Explaining other objects would probably only be possible under great circumstances, such as explaining blood cells, or not possible at all if it was not something you could see with the naked eye - or if it was too abstract. How would one describe the circumference of the earth, if the knowledge about planets, geometrical bodies, and the measures in which the circumference of the earth would be given, is not known? Last but not least, one would need numbers and letters, as well as an understanding of arithmetic and reading, and the knowledge of language in general. A subject which deals purely only with abstract concepts is *mathematics*.

Mathematics is a very abstract and theoretical subject, and the concepts and ideas involved are difficult for many people to understand. Its theoretical nature makes its practical application even more complex. Now, unfortunately, many people think they don't understand mathematics. Lynn Arthur Steen refers to this as "mathophobia"[149]. This fear is said to be based on the teaching style of draconian teachers who force students to memorize mumbo jumbo that has nothing in common with the actual understanding of mathematics.

In retrospect, I can tell you from my own experience that the connections between the individual areas should have been made more visible. For example, the fact that a function or a figure is a relation between two quantities only became clear to me during my studies. In general, in my opinion,

---

[149] Steen, 1978

many concepts of mathematics were only touched upon and not shown in connection with all the others. I'm sure there are reasons for this, especially since children would then naturally want to understand the other concepts as well, and parents would wonder why these concepts were mentioned in school when they had little relevance to the end of the school year. I can understand that in school you have to limit yourself to superficial knowledge of the individual areas, because you are trying to educate children about different sciences in a comprehensive framework.

For me, however, this had a rather negative effect, as I lacked context. Today I always learn a bit more, find out more details for myself, understand why things are the way they are, and then learn what I need to learn for the test. But for many people, that's too much effort – they only want to learn what they need. This is a pragmatic approach that I can understand, but if one were to understand the benefits of more detailed autodidactic study, one would soon realize that the additional effort of such an endeavor would not be in proportion to the return one would get from it. Returning to mathematics, if one had perhaps also explained in concepts and their interconnections, one would understand more complex topics much more easily. For example, advanced concepts in mathematics include topics such as calculus, algebraic geometry, and topology. These concepts can be challenging even for experienced mathematicians and require a high level of abstraction. In addition, many complex concepts in mathematics are interrelated (system) and may involve multiple levels of abstraction or complexity. Would you know what integral calculus or differential calculus is used for? Only a few know it, but have learned to calculate with it. This also prevents interest in a subject from building up. If there is no practical relevance, i.e. if it is not explained what these things can be used for, then the interest is soon gone. And if there is no interest, then you don't focus either, so you take very little with you. If you have to deal with an abstract concept anyway, which is completely taken out of context, and there is no focus, then you soon don't understand anything anymore. And since later topics build on earlier ones, mathematics doesn't get easier either, but harder and harder if you haven't understood the basis.

Now it is not always the case in mathematics that there must be a practical application. Pure mathematics deals with topics that have no practical application. If you haven't understood that and keep trying to establish a reference to something that doesn't exist, you'll have an even harder time. So how can you continue to fuel interest in mathematics? By understanding that the beauty of mathematical formulations is not made to be applied always and everywhere, but that they stand on their own, one can use them to calculate something in abstract form that goes beyond what we currently understand in our reality, which will someday allow us to transform these calculations back into applications that are close to reality. Because if it doesn't work in mathematics yet, then we can't build it yet...

"Mathematics doesn't need an application. Mathematics is on its own."[150]

---

[150] In an introductory course on physics, Mag. Hans Frost-Schimpf once said "*Mathematik braucht keine Anwendung. Mathematik ist für sich alleine.*"

It is difficult to say that a particular branch of mathematics - namely, pure mathematics - has no practical applications, since mathematics is a fundamental part of many fields and is used to solve a wide range of problems. Mathematics is used to model and understand various phenomena, and many mathematical concepts and techniques have been developed over the years to solve practical problems in a variety of fields. Pure mathematics, for example, is a branch of mathematics that focuses on the abstract study of mathematical concepts without necessarily considering their applications in the real world. This includes areas such as algebra, geometry, and topology, which do not have as many direct practical applications. If you consider to improve your logical thinking, I would advise you to focus on pure mathematics. It can also have applications, in some cases, as it provides a foundation for other branches of mathematics and leads to the development of new mathematical concepts and techniques that are used to solve practical problems. But that is not primarily the reason why one would need to study it – in fact, its beauty lies in abstraction even without having any practical applications. The study of complex concepts in mathematics is in any way useful and can be a rewarding and challenging task. It can provide valuable insights and perspectives on a variety of phenomena.

Learning new concepts can help you expand your knowledge and understanding of the world and develop new skills and abilities. It can also help you think more critically and creatively and solve problems more effectively.

When learning complex topics, you might wonder what you need them for. After all, they are time-consuming to learn. Why should one take on this effort? What do *I* need this for, why am *I* learning this? That's a question I asked myself as a child quite often. Today I think completely different about it when learning something complex. As individuals, we often question the value of the knowledge we acquire and its relevance to our future lives. Our perception of what we "need" to know is often based on past experiences and is not always applicable to our present or future self. Therefore, it is crucial to develop a broad foundation of knowledge and skills while remaining open to new learning opportunities, rather than just focusing on what we think we might need in the future. While you might not need a lot of things you learn today, you also did not learn a lot of things that might have been beneficial later in your life. Instead of asking on how this complex topic you are learning today might be useful later, believe that it will be and that the process of learning it is more important than knowing the thing you are learning about.

Learning is a continuous process, and it is impossible to predict what skills or knowledge will be essential in the future. Building a diverse knowledge base ensures that we are better equipped to face unforeseen challenges and opportunities. At the same time, it is essential to prioritize and focus on learning that is relevant to our current situation. This requires a balance between a broad base of knowledge and a targeted approach to learning. Should we aim for specialization or on gaining universal knowledge in many different fields? In the book *Range: Why Generalists Triumph in a Specialized World*, David L. Epstein argues that a broad field of knowledge generally helps us draw from a large pool of information and use analogies to solve problems more easily. The more concepts you learn, the more analogies you can make, and the easier it is to learn new concepts.

Learning what we need today is a very short-term view of things, the world, and life. How can we know what our life will be like in 10 years? If you read a page a day starting today, you will be one step ahead in 10 years. But we all know we could read faster, yet we focus way too much on what we can accomplish in a short amount of time. We focus too much on what we can achieve in a day and underestimate what we can achieve in a decade. Sometimes, not needing something can be beneficial, and learning for the sake of personal growth and development can be more valuable than learning solely for practical reasons.

> "What becoming a long-term thinker most requires is character. It's the courage to carve your own path, without the reassurance of doing exactly what everyone else in the crowd is."[151]

A bachelor's degree takes 3 years, a master's degree 2 years. You can read a 300-page book in a month, and learn the basics of a language in 3 months.

In 10 years, you could read 120 books and thus already become an expert in a field - provided you want to. Or you could just read for the joy of it. Probably that's the best tip. Why worry about how long something takes? Time passes anyway. If you enjoyed it, isn't it irrelevant how much time it took? That's how I feel about the Chinese language. I probably won't become a sinologist and my linguistic skills in

[151] Dorie Clark, *The Long Game*

Chinese won't compare to a native speaker. But then who cares how well I can speak Chinese if I had fun learning it?

> „And most people say, 'Well, I don't have time to read outside my field.' I say, 'No, you do have time, it's far more important.' Your world becomes a bigger world, and maybe there's a moment in which you make connections."[152]

Whether I need it or not should not be the motivation. In the end, whether we might need it or not is actually irrelevant. It is better to have the knowledge and not need it than not to have it and need it. Focusing on immediate needs is a short-term view of learning that disregards the long-term benefits of personal growth and development.

> "The more interested students are in a subject, the more involved they become in their assignments, putting effort into their studies and engaging in deeper levels of thinking. Experts believe that increased student engagement in math and science at school will eventually lead to involvement in math- and science-related after-school activities and career aspirations."[153]

Reading books about general knowledge just to know what is generally known is not something I think is useful. Such books are based on factual knowledge, not on understanding new concepts. The second would be much more important in order to understand and make sense of the world. It is more important to understand *why* something is rather than knowing *how* it is.

By questioning why something is the way it is, you learn to understand *how* it is and *how it came to be* the way it is.

---

[152] David L. Epstein, *Range: Why Generalists Triumph in a Specialized World*
[153] May et al., 2022

Why something is the way it is, is therefore more important than simply learning that something is the way it is. Asking why-questions over and over again really helps to understand things. It's what children do naturally, but somehow, many people lost that curiosity. Knowing *why* something *is* also opens up the possibility of questioning one's own view, while knowing *that* something *is* the way it is does not. The why *questions*, the other takes the fact as given. By questioning and finding out more just because out of curiosity, something else happens that is incredibly important: you learn to become more curious and understand that it's important to question other things as well.

We cannot accurately predict what our future will hold, and therefore, it is crucial to adopt a long-term perspective towards learning. It's a never-ending process. Even small efforts, such as reading a page a day, can accumulate to significant knowledge acquisition over time. As such, it is essential to consider the value of learning beyond its immediate application and value its intrinsic worth. Prioritize what is relevant to our current situation, but appreciate the intrinsic value of learning more beyond our current situation and need. By having a long-term perspective towards learning, we can accumulate significant knowledge and skills that will be beneficial to us in unforeseen ways.

So what is *the* knowledge that is important? While pursuit of knowledge is a lifelong journey that requires dedication, perseverance, and curiosity, there is not necessarily *the* important knowledge but rather the act of *pursuit of knowledge* that is important. The relevance of acquiring knowledge cannot be overstated, but it is not always easy to explain *why* a *certain* knowledge is important. We are often motivated by the pursuit of a better job, but a university degree opens up many more possibilities beyond career advancement. It allows us to explore a broader field of interest, which in turn paves the way for new interests, motivations, and curiosity. This broader area of interest then paves the way for new interests – e.g., because one realizes the importance of said interests –, which in turn creates a new motivation and curiosity. Once we reach a barrier of understanding, we simply want to find out more and *go beyond simplicity*.

But explaining this curiosity to others who have not yet been exposed to such theories can be challenging, as they may not understand why one is so interested. This is because they have not yet experienced the importance of the knowledge we have acquired. As scientists, we have a responsibility to communicate new knowledge and theories to the public, and this is where writing a book like this can be a valuable contribution. We can still communicate our enthusiasm and share our knowledge with others in a way that they can understand. This not only benefits others but also reinforces our own understanding of the subject matter. By explaining what has already been done in a simple yet informative manner, we can create a greater understanding and appreciation for our field of study. We should strive to avoid oversimplification while still presenting the information in a way that is accessible to a wide audience.

One could argue that studying complex topics at university is not necessary to become successful financially, such a viewpoint is limited in its scope, taking into account only the immediate future, and neglecting the long-term implications of such a stance. One who subscribes to this ideology will be unlikely to devote their attention to intricate and novel subjects that may not have an immediate financial reward. While advanced mathematics, for example, has immense value, many people will not see it – because they do not understand it. This capitalist belief, that is held by many individuals – that something holds value only if it can generate monetary gain – is a perspective that warrants contemplation. This notion is reflective of a larger cultural issue, whereby society is inclined to evaluate worth solely in terms of monetary compensation. This perspective often leads to a disregard for the value of intangible assets, such as creativity, innovation, and intellectual property. The enthusiasm for this subject cannot be explained to someone who does not yet understand it. Therefore, all you need to do is take the time and learn it to understand other people's enthusiasm for it. One must invest the time and effort necessary to learn and grasp the nuances of the topic – and maybe get a better understanding why people love these nuances. It is through this acquisition of knowledge that one can begin to develop the motivation and enthusiasm required to pursue the subject with greater fervor, and then gaining more insights about it. By studying the things we do not yet understand, we can appreciate the importance of knowledge – and inspire others to do the same. While we may not be able to master every complex subject, we could at least strive to broaden our understanding and knowledge and build a solid foundation in order to understand the importance in a bigger picture of it. While it may be easy to live without knowledge that we do not understand yet, we only understand its importance when we take the time to study it. It is through a dedicated pursuit of knowledge that we can hope to fully appreciate the complexities of the world we inhabit. We need to *expose* ourselves to them.

References to mathematics and physics are made very often in this book. If you enjoy working with abstract concepts and ideas and have a talent for mathematics, studying abstract mathematics could be a rewarding and fulfilling experience for you. Mathematics can help you develop critical thinking and problem-solving skills, and can also be valuable in expanding your knowledge and understanding in many different fields such as science and engineering. On the other hand, if you don't have a strong interest in mathematics or don't like working with abstract concepts, studying pure mathematics may not be the best choice for you. It is important to choose subjects that you are passionate about and that will help you achieve your goals. It is difficult to say which subject area has the most abstract concepts because different people have different opinions about what an abstract concept is. Subjects that deal with theoretical or philosophical ideas, such as mathematics, physics, and philosophy, are thought to have the most abstract concepts. These subjects often study ideas that cannot be directly observed or measured, such as space, time, and consciousness. Therefore, understanding them often requires a great deal of creativity and imagination.

Abstract mathematics involves working with abstract concepts and ideas, which requires a high degree of mental flexibility and the ability to think logically and systematically. By learning abstract mathematics, individuals can improve their ability to think critically and solve complex problems - skills that are useful in many different fields. It is especially important for advancing scientific and technical knowledge.

Many areas of science and technology, such as computer science, physics, and engineering, rely heavily on abstract mathematics. Those who learn abstract mathematics can better understand and contribute to these fields, which are critical to our understanding of the world and to technological progress. Mathematics is also important for its own sake. Many people find mathematics beautiful and fascinating in its own right and enjoy studying it for the intellectual challenge and sense of accomplishment it provides. Learning abstract mathematics can be a rewarding and fulfilling experience for those who are passionate about it. The ability to think critically, expand scientific and technological knowledge, and experience intellectual challenge and fulfillment are all reasons why mastering mathematics is essential.

If you are not familiar with the terminology used to describe a new concept, it can be difficult to understand the concept itself. Therefore, it is helpful to first get a basic understanding of the concept without worrying too much about the specific terms used to describe it. You can do this by reading or listening to an overview or explanation of the concept, and by using examples or illustrations to help you understand the main ideas. Once you have a general understanding of the concept, you can focus on learning the specific terminology associated with it. This may require looking up definitions of key terms or asking others more familiar with the terminology for clarification. By breaking the concept down into smaller, more manageable pieces and seeking additional resources and support, you can gradually build your understanding of the concept, including the terminology used to describe it. Learning new concepts can keep your mind active and engaged and help you stay curious and open to new ideas and experiences, understand the world around us and interpret and comprehend the various phenomena, events, and experiences we encounter in our lives, navigate daily life, make informed decisions, and adapt to changing circumstances. There are probably many ways to improve abstract thinking, and you may not necessarily agree one hundred percent with what I write below. Nevertheless, these are the best methods that have improved my abstract thinking and thus my understanding of everything related to it:

Programming and Computer Science: Understanding concepts in programming and applying them in different programming languages is a surefire way to improve your abstract thinking. Understand the art of programming as well as the technology behind it, software architecture, and the computer itself.

Writing and Reading: The act of thinking on paper makes thinking easier. By explaining things clearly to others, you make sure you understand. It also helps others learn what you've learned, which is pretty cool in itself. Plus, you'll run into barriers in your own language that make it difficult to explain the things you want to explain. If you read more, you can write better - and understand everything in between. There is no book or website that I would recommend. You should focus on what you don't yet understand and read as much as you can. I'm still looking for the most difficult book ever written....

Exploring Sciences: Having a passion for science itself and questioning everything you read is a good way to improve your overall thinking. If you don't understand something yet, it might be useful to learn it, as it could help you in your future endeavors. In the past, I referred to this as *expanding your awareness.* Only when you have the awareness that you should learn certain things will you learn them. But if you live a simple life and never deal with higher concepts, you will never enjoy such awareness either.

> „Learning is hard work, but everything you learn
> is yours and will make subsequent learning easier."[154]

Abstract thinking enables us to think about complex relationships, recognize patterns, solve problems, and use creativity. Although we cannot know today what we will need to know tomorrow, abstract thinking will enhance our understanding of the world, the universe, and life by enabling us to analyze and understand complex concepts and ideas that are not directly related to our concrete life experiences. This helps us to recognize and understand patterns and connections – which might otherwise remain hidden.

We have learned what concepts are and that they represent the real world[155] – our reality. If we want to describe the world, we need concepts to describe it. Also, if we want to describe more than the phenomena of reality, namely concepts like mathematical theorems, we need the corresponding concepts. Learning in concepts is nothing else than understanding these concepts on the one hand and showing

---

[154] Haverbeke, 2018

[155] The one which do as well as the ones which do not.

them in an understandable context on the other hand. In the book *Erfolgreich lernen mit ADHS: Der praktische Ratgeber für Eltern*, the authors provide some interesting points on how to improve learning:

> "[Good students] don't just read these books. They read them, think about what is written, network them with their prior knowledge, immerse themselves in intense imaginings, look up uncertainties or unanswered questions, and automatically repeat by reading exciting passages several times or telling someone else about them. In addition, children instinctively try to make interesting information "tangible." They draw pictures on their favorite topic, make models, or act out certain passages of a story with play figures or live with a friend. So they're doing exactly what leads to strong, lasting connections and good networking in their brains."[156]

By connecting the concepts together (*system*) you can better understand as well as ultimately remember them. Practice actively engaging with the material by asking questions, participating in discussions, and trying to apply what you learn in real-world contexts. Likewise, it would be possible to actively use these concepts by writing a book. Review and reinforce what you have learned regularly by reviewing your notes, summarizing key points, or discussing the material with others.

---

[156] Translated from: „*Diese Bücher lesen sie nicht einfach nur. Sie lesen sie, denken über das Geschriebene nach, vernetzen es mit ihrem Vorwissen, tauchen in intensive Vorstellungen ein, schauen bei Unsicherheiten oder offenen Fragen nach und wiederholen automatisch, indem sie spannende Passagen mehrmals lesen oder jemand anderem davon erzählen. Darüber hinaus versuchen Kinder instinktiv, interessante Informationen „greifbar" zu machen. Sie malen Bilder zu ihrem Lieblingsthema, fertigen Modelle an oder spielen bestimmte Passagen einer Geschichte mit Spielfiguren oder live mit einer Freundin nach. Sie tun also genau das, was zu starken, dauerhaften Verbindungen und einer guten Vernetzung in ihrem Gehirn führt.*" – Rietzler & Grolimund, 2023

Creating:
Concepts and Systems

———

We are now able to think in abstractions, i.e. in concepts and systems. What if the current concepts and systems do not provide enough to communicate everything we want to understand? Let this be an additional chapter within the third part of this book – a *challenge* to create our own *concepts* and *systems* to better understand the world and life itself.

Did the *term* or the *concept* – the idea behind it – come first? Every word describes something and what it describes must either pre-exist or be invented. The *term* describes the *concept* in such a way that we need a single word – or a few words – to communicate an idea. It is possible to create our own *concepts* for abstract ideas. In fact, people do this all the time. Whenever we think about something that has no concrete physical form, such as love, happiness, or justice, we create a *concept* in our minds. In doing so, we use our past experiences and knowledge to form a mental representation of the idea, which we can then use to understand and think about it. Geniuses have likewise used new terms to abstract their concepts. The simplest kind of new terms are *categorizations*. An example would be the types of books (children's books, non-fiction, novels, etc.) according to their *content*, target *audience*, and so on. Another possibility would be by their *format* or *texture*, so there are hardcovers or paperbacks, or by the number of pages, like a *pamphlet* or a *tome*. Although the second term is colloquial, we have a clear idea of what is meant.

A term can also be used to describe the relationship between each other – in a *system*. Not everything can be categorized into a system for ease of understanding. Some phenomena, especially those that are very complex may not fit into an existing system and it may be necessary to develop new systems or frameworks to better communicate these concepts and their relation to each other. Thus, for each kind of relationship a new *system* could be created and a corresponding word could be introduced. Geniuses created their own systems because they provide a clear and organized way to understand and examine complex information. By grouping similar concepts or objects and classifying them according to certain criteria, systems make it easier to understand the relationships between them and to recognize patterns and trends. In this section, we will now note how we can create such systems ourselves. But first, let's look at existing systems to elaborate on the benefits of this.

The periodic table of elements, for example, classifies elements according to their atomic structure and properties and is a useful tool for understanding the behavior of different elements in chemical reactions. Similarly, Linnaean taxonomy classifies living organisms into a hierarchical classification system based on their physical and behavioral characteristics, facilitating understanding of the relationships between different species and their evolution. In addition, the use of systems in science and research also facilitates communication and knowledge sharing with others. For example, the Dewey Decimal System, the SI System, and the ICD make it easier for researchers to find and access relevant materials and for physicians to diagnose and treat patients.

These systems help to make the organization of science and research more efficient and understandable, and allow easy comparison between different fields. This also reduces the time and effort required to find and understand relevant information. In addition, systems facilitate the understanding of complex information by providing a clear and concise structure. Geniuses created such systems primarily because they wanted to order these subject areas for themselves. The order character was essential for geniuses - they wanted to order the knowledge of the world. In the meantime, one restricts oneself to creating systems in different sciences so that at least the knowledge within these sciences is ordered. But how do you create *systems*? There are different ways to categorize knowledge, depending on the purpose and context:

1. Determine what type of knowledge you want to organize and why. Consider the audience, format, and context of the knowledge.

2. Gather all the information you want to include in your system and review it to ensure it is accurate and relevant.

3. Identify the key concepts and relationships that will be used to classify and organize the knowledge. For example, you may want to organize the knowledge by topic, level of abstraction, or format.

4. Using the criteria you have developed, create a structure for organizing the knowledge. This may be a hierarchical system, such as a tree structure, or an

> interconnected system, such as a network of interconnected concepts.

5. Insert the collected and reviewed information into the system, following the structure and criteria you have developed.

6. Test the system, using it to find and understand information, and refine it as needed.

7. Create documentation explaining the system, the criteria used, and how it works to help others understand and use the system.

8. Review and update the system regularly to ensure it remains accurate and relevant.

It does not always make sense to use a new word if one does not want to focus on a specific form of the thing. Therefore, when speaking of books in general, it would be easier not to enumerate every single word describing the different types of books (non-fiction, novels, ...), but to use the corresponding general term (*books*). We now know the constitutions of a concept as a form of specialization – just by type, nature and other properties - and as a form of generalization. These possibilities enable us to design new terms to describe present phenomena as concepts. To do this requires a new way of looking at things. For example, the mathematician who calculates in the decimal system may consider the calculation $1 + 1 = 1$ to be wrong. Also for the computer scientist who calculates in the binary system it would be wrong – because for him the answer would be 10. The logician, however, would consider $1 + 1 = 1$ a true statement[157]. We recognize therefore that a different view on the same phenomenon brings different results. Accordingly, it is not reprehensible to design a completely new view on already existing concepts.

It would make little sense to arbitrarily select concepts and give them a new meaning. In order to define a new concept, it must first be determined whether there is a need for this concept. Where does this need come from? For example, if there is a concept that describes a thing, a situation, or something else, but it turns out that there should be a separate word for particular situations - to make it clear that it is precisely that situation where this concept applies - it makes sense to introduce a new word. It is usually not a good idea to create separate words and concepts without a clear purpose or context. In most cases, it is better to use existing words and concepts that are already well defined and understood by others - but we know that this is not always possible. Clear terms allow for clear and effective communication and avoid confusion or misunderstanding. In some cases, however, it may be necessary or appropriate to create new words or concepts to better express an idea or concept that is not adequately captured by existing terms. How does such a situation arise? Sometimes it can be difficult to describe complex or abstract ideas using existing concepts. In these cases, you may need to define new concepts to convey your thoughts more accurately and clearly - which is why many geniuses have defined their own concepts. Some of them have even created a new field of study. If you are working on a new area of research or study, you may need to

---

[157] *Logical conjunction* used in mathematics, logic, and linguistics.

define new terms to clearly describe your work and distinguish it from other areas. And that's exactly what geniuses did! But not for the pure motivation of being different. Sometimes there was no other way than to define new words or use existing words in a new context to convey the ideas that were important to them.

It is no longer enough to be good at just one thing. You have better chances in life if you have dealt with many things.[158] Broadening one's knowledge base and gaining expertise in diverse domains can prove invaluable in many areas of life. With a solid foundation in multiple disciplines, individuals can gain an advantage in furthering their education and exploring new avenues for professional and personal growth. The world of work is changing and the winner is the one who adapts.

> "The future belongs to those who learn
> more skills and combine them in creative ways."[159]

As the world continues to evolve, there is one hobby that remains timeless: learning. It is a hobby that will never be obsolete, and its importance only continues to grow. In today's fast-paced world, where digital entertainment dominates our leisure time, it is crucial to make time for activities that nourish the mind and expand our knowledge. As an avid reader, I always strive to learn something new and thought-provoking through the books I read. However, I found myself in a rut, consistently receiving book recommendations for popular bestsellers that failed to expose me to new ideas. This

---

[158] Ditta et al., 2020
[159] Greene, 2013

was a significant flaw in the recommendations system, and I wanted to find a way to discover more captivating books and also learn more about difficult complex concepts.

I came across a video[160] that provided recommendations for books on math which fascinated me. The video contained a list of both old and new books, some of which were out of print and hard to get. Nevertheless, I purchased many of them, even if I wasn't entirely sure I would read them right away. I recognized that books such as these were what I wanted to spend my time on. Then something happened… By the time I was coming back to the online bookstore and wanted to look around, the book recommendations I received were quite different. I discovered more books that piqued my interest, and the more I browsed, the more recommendations I received. This was a win-win situation for me, as I was able to explore new areas of interest, and the algorithm became more tailored to my preferences.

In life, it's important to focus on the complex and venture where the majority won't. This is equally true for reading. It's easy to become comfortable with the books we read, but it's important to explore new genres and styles to gain a more diverse perspective. However, it's essential to remember that purchasing books is just the first step; reading them is equally important. It's not enough to have a collection of books; you must engage with them to truly gain insights and expand your knowledge.

I often encourage readers to create a personal *vision* of themselves, an image of the person they aspire to be. But, it's

---

[160] The Math Sorcerer | Learn Mathematics from START to FINISH (2nd Edition): https://www.youtube.com/watch?v=didXE0HkSC8

equally important to act like that person today – by purchasing books that align with the vision of yourself, you can fill your library with new ideas and change your environment accordingly. This will help you become the version of yourself that you want to become.

Indeed, the pursuit of knowledge and learning should be a lifelong endeavor. However, as we grow older, we often lose the childlike curiosity that once motivated us to explore new subjects and ideas. To regain this sense of wonder, we must actively seek out the things that excite us and keep our passion for learning alive. Reading offers an opportunity to immerse oneself in a world of knowledge, imagination, and creativity. Whether it's a novel, a biography, or a non-fiction book on a specific topic, the act of reading allows us to expand our horizons and explore new ideas.

Another great hobby I pursue is learning new languages. Not only does it provide a practical skill, but it also opens up a whole new world of literature and culture. For me, it has always been the sound and complexity of a language that fascinated me. My fascination with Mandarin Chinese, in particular, stemmed from its reputation as one of the most challenging languages to learn[161]. Then about ten years ago, I came across a book about polyglots.[162] During that time period, I lacked the awareness of the term 'polyglot.' However, when I discovered the concept, it left a profound impression on me and instilled a strong motivation to pursue language learning of more than a single language. The newfound knowledge liberated me from the confines of

---

[161] While Mandarin is considered a difficult language, there are other languages, such as click languages, that are even more challenging

[162] Erard, 2012

limiting myself to one language and opened myself up to learn many more – 60 to be exact. Of course, I was never able to speak all the languages at the same level, but at least I gained a first impression of them and built myself a foundation. Another thing happened, that was much more interesting than actually being able to speak the languages I learned: My learning methods improved. Not only that, but my confidence also grew exponentially. It can therefore be said, that the sense of accomplishment that comes with understanding something that once seemed impossible is unparalleled. It was a great confidence booster, and it can also foster a sense of curiosity and a desire to continue learning, leading to a lifetime of personal and intellectual growth.

> I appreciate you think you know what's going on. That you see yourself in some position of involvement and responsibility. But what you know of all this doesn't even begin to scratch the surface."
> – Linda Holland[163]

In order to effectively navigate through complex, confusing, and difficult concepts, it is important to develop a range of intellectual skills and strategies. The intellectual skills can arise before learning or with learning something complex. This means that – in any case – it is useful to learn something new and complex, because it gives us the ability to learn more in the future. Intelligence is not a fixed value that can be measured once, which then tells you that you can or cannot do this or that. Furthermore, past experiences are no barrier to learning certain things in the future. The fact that you didn't understand something doesn't mean that you wouldn't understand it if you'd spent more time with it later on. It just means that you didn't understand it at that point in time. But understanding something that is difficult to understand is a lot of work and too much effort for many people. I hereby want to show you that it is worth the effort and also that it becomes easier over time. It will always be laborious and it will always be a challenge because we are constantly learning and exposing ourselves with new concepts. But it will become easier over time because we can use analogies from the past and we also have more confidence in ourselves. After all, we have done it before, so why shouldn't we be able to do it again? Now there are some advantages from this endeavor, such as developing a deep understanding of underlying principles and concepts, engaging in critical thinking and analysis, seeking out additional information or resources, and improving your abstract thinking skills. Abstract thinking is so important to me, in fact, that I will devote a separate chapter to it in this book. By grappling with complex, confusing, and difficult concepts, we can expand our understanding of the world, develop new skills and competencies, and cultivate a sense of intellectual curiosity and wonder. It is indeed the case that as new complex concepts are learned, the interest and curiosity to learn new such concepts increases. Furthermore, it is important to recognize that intellectual challenges like learning new complex concepts can also offer opportunities for growth and development in other areas of ones life. What we want to do is to understand these concepts, to recall them, and to actively use them in whatever is interesting to us. The ability to recall and apply these concepts is a hallmark of intellectual growth. So just reading them is not enough. However, I am aware that many complex concepts do not have immediate application in our personal lives. This means that the sense of learning such concepts might not visible to us, and therefore may not be possible to grasp. This also means that the motivation to learn such concepts could be very low to some people – hence which is why I have written this book and why you are reading it. A lot of people have asked me what skills they should learn to increase their value in the marketplace. That's the wrong question. Find something that really excites you. Not because it makes you money, but because you love the subject itself. When it comes to foreign languages, people often ask me what language they should learn. They think of Spanish or French because it will help them get a better job. But that's not the importance to consider. Learn a language that is so exciting to you that you want to spend a long time with it, because you will have to if you really want to learn it. Don't worry about this skill having "value" in the marketplace. I have not only found a lot of joy in the process of learning, but also improved my ability to learn by focusing on what really peeks my interest – and not on focusing what has the most value in the marketplace. This interest in different languages has not only motivated me to learn more, but also for the joy of learning itself. As a result, I have developed a strong ability to learn more effectively. By learning things that interested me, I was better able to engage in the learning process. Through the process of learning multiple languages, I have gained a wealth of knowledge and skills that go far beyond language learning itself. I have learned to approach the learning process more strategically and efficiently and have developed a number of techniques and methods that have enabled me to progress more efficiently in my topic of interest. It came to me intuitively.

> "Through their learning activities, students' worldviews and thinking are formed, and their conscious attitude to the social system is formed. It is known that in the process of learning, the student's activity increases and his interest in academic subjects is formed."[164]

My personal joy and interests has enabled me to progress in my studies with greater ease and efficiency. That is, not only in the studying of languages, but also in the study of computer science, mathematics, engineering, and so on. And it all started with a personal interest rather than a focus on practical applications. Almost a decade ago, I completed my first Bachelor's degree in economics, despite having little interest in the subject. While I found it challenging, I was determined to finish and did so. Later, I pursued a Master's degree in the same branch, but found the lectures rather disappointing. Eventually, I embarked on a second bachelor's degree in business informatics. I am now on the verge of graduation and contemplating another bachelor's degree, but this time it would be in business engineering. This would prepare me for a master's degree in space engineering – something that I too would want to focus on out of my personal interest. Do I need them for my job or for work? Not necessarily and I have not thought about pursuing either professionally – that is, getting paid for doing engineering. Of course, math is still important for me too as I will be needing it for engineering, but also because of my personal interest in the field of topology. This means, I have been studying it part

[164] Renatovna & Renatovna, 2021

time while working full time. While pursuing a Bachelor's degree is not necessarily the right choice for everyone, taking calculated risks to learn something new and expand one's knowledge base is crucial. Thus, it is essential to focus on learning something new, even if it is complex or difficult, and to keep learning throughout one's life. In retrospect, having a Bachelor's degree in a field that did not pique my interest meant that pursuing a Master's in economics was not the right decision. It took some time and effort to overcome my hesitation, but at the end, I have taken the risk and stopped pursuing it. At that time, I did not know if I would ever go to university again. Learning and studying for only one job or one type of work and doing that for the rest of your life is very unlikely. Should you go to university or not? That is not the point of this chapter. Think of it more as why you should study something new in general – maybe focus on new and exciting technologies. But above all, you should deal with what you don't yet understand. You should *go beyond the simplicity* of everyday life and focus on something complex. Learning the complex is not always easy, but it is a worthwhile endeavor. Many people shy away from tackling difficult subjects, preferring instead to stick with what they know or what is familiar. But those who are willing to take on the challenge of learning something complex will reap significant rewards! By learning something complex, you are developing a skill that many others will not possess – truly, it's lonely at the top. This can give you an edge over others in your field when it comes to finding a job or advancing in your career. Of course, this is not what matters most of all, but it remains a crucial factor to consider. Acquiring new knowledge has become more accessible than ever. The internet has made it possible to access a plethora of information sources, including the widely popular online encyclopedia, Wikipedia. This has really revolutionized the process of learning, allowing everyone with internet access to acquire new knowledge with ease. The large quantity and diversity of information available on the internet makes it an ideal medium for exploring new subjects and expanding one's understanding of various complex topics. Despite its advantages, learning through online sources such as Wikipedia requires a certain level of caution. Of course, it is – as many others – a biased source of information. It is crucial to evaluate sources carefully, verifying the accuracy and credibility of the information presented. Moreover, because the internet allows anyone to contribute content, some articles may contain errors, bias, or misleading information. It is recommended to place one's attention on areas of lack of understanding. This necessitates a willingness to engage in introspection and receptiveness to new information. Wikipedia only provides a surface-level overview of a topic. To gain a comprehensive understanding, further research is necessary, such as watching YouTube tutorials or delving into academic literature. This requires a commitment to studying the subject matter beyond the superficial level and embracing the complexities inherent to it. Amidst the vast expanse of knowledge yet to be uncovered, it is all too common to be captivated by the countless prospects and possibilities that await. Nevertheless, it is of paramount importance to fixate one's focus and engage in prolonged periods of concentration. Only then can we fully immerse in the task at hand and realize the truly remarkable potential of our pursuits. You really have to *go beyond simplicity*.

Applications

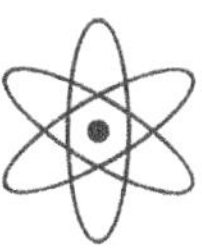

Making Use of What You Learned

"Unless you challenge yourself, you will not learn. You will
not grow.
Education is the art of challenging and overcoming."
Amit Ray

Applications in Everyday Life

"Without electronics, our world would grind to a halt."[165]

Now, the question naturally arises as to when one could actually apply what has been learned and whether an opportunity will ever arise for that. The answer is that this is not the case that such opportunities ever arise, unless of course we actively deal with complex concepts. We are aware that it is difficult to capture all the information, as this is hardly possible due to the abundant complexity. There is no situation in everyday life where one would need complicated and complex concepts. One must actively choose to engage with them.

Therefore, the chapter on *Application* tries to establish what *activities* one should engage with. The subsequent chapters aim to explore the purpose of applying what has been learned in this book – one of many attempts to communicate something in a way in which it has not been communicated before. Another book, which has been fully written to correspond to its scope and, hopefully, provides the reader with sufficient information about how it behaves in its peculiar sense. Let's be clear: there are practical use cases of learning and understanding complex concepts, but these are mostly beneficial to a small minority of our society – that is,

---

[165] University of York, 2023

when you are working in a company that deals with complex concepts, as found in science and engineering.

When it comes to everyday life, we have to dig a little deeper and ask questions which we would normally not ask ourselves – since we do not need an answer to them to survive. That is to say, we actively need to spend our time with media that deals with complex concepts in order to get to such questions. Let us look at a few examples on how we can deal with complex concepts in everyday life – we will look at each part in depth in the chapters that follow:

* *Career:* Understand that technologies will continue to influence your job in the future, choose a job where you deal with complex topics rather than simple ones, understand that dealing with complex concepts today will help you in any other job and in your life as well.
* *Books:* Choose complex books over others, change the algorithm when shopping online, focus on reading slow and steady rather than speedy, acquire books you do not understand yet.
* *Education:* Different forms of education will expose you to complex concepts, dealing with them as part of a degree program will help you to see the connections, teach others about what you learned to remember much more.
* *Output:* Creating and managing projects and learning more about complex concepts in such an endeavor.

The question still remains as to how important a study of complex concepts is going to be in the future. What do you you need this for if societal, economic, and technological shifts are likely to reshape the entire landscape, potentially leading to the emergence of entirely new systems and entities beyond our current comprehension. In addition, all of this study of complex concepts takes up a lot of time, and naturally, one also expects a return on investment – they want to earn money with it. In many cases, that might not necessarily be guaranteed if one were to make a career change as you would have to start working from the ground up. I question the practical use here and I advise you to do the same. It is one thing to be excited about the complexity of certain topics – which I am and I am investing a lot of time and money into this endeavor – but it is another to actually see the practicability of it. I can understand if learning complex concepts seems pointless to you today. The chapter on *Applications* is not about you finally being able to apply your knowledge about topology or rocket propulsion in everyday life, but rather about getting exposed to such concepts, how you will meet them in everyday life – through books, higher education, and sometimes your job – and how you will then understand the beauty of learning them. There are unwritten rules on how to live a life. One of them is choosing a job and sticking to it for a longer period of time – maybe even for the rest of your life. This sadly takes away from the available wealth that a life could hold – the incredible variety and range of things one could learn if they only took the time to learn them. An important skill, the significance of which only becomes apparent when you use it, is the ability to make good decisions. Therefore, we must be capable of predicting the future to some extent and, based on the current information available to us, make a decision that will support us in the best possible way in our future. Naturally, it is not possible to predict the future, and there is no definitive best decision. However, what we can do is learn to estimate what could be. Understanding macroeconomic trends is only possible when we have grasped the complex connections, which is why it is important to learn complex concepts in order to acquire this very ability. I have noticed that people sometimes make wrong decisions even in the simplest of choices. The reasons for this are numerous, and I do not wish to delve into them. Let's provide a simple example to explain this situation: Someone is asked if they want to have a Ferrari. Many people would answer this question with a yes because we assume that it is an object that many people desire. However, and this is easily forgotten, owning a Ferrari comes with many responsibilities that the owner must attend to on a monthly basis. These include insurance costs, fuel expenses, possible repairs, parking, and several other things. If people are made aware of all these aspects, they would probably answer the question of whether they want to have a Ferrari differently. It is not that something has changed about this Ferrari, but rather that these people now have more information about the situation. The information was always available; they simply did not inform themselves. We could say they would have made a wrong decision initially if they were to change their mind after such a short time. The lesson in the complexity of certain objects, items, and situations enables us to consider all these things. So, instead of answering a question about whether they want something with a simple yes, one will first weigh what else it entails.

> "New problems are usually first approached using established tools, whether or not it is clear that they are appropriate, because it is all we can do at the time."[166]

---

[166] Krauss, 2007

Applications:
Personal Career

"It is a sad fact that from early childhood we are tyrannized by the moral myth that it is right, proper and good to leap out of bed the moment we wake in order to set about some useful work as quickly and cheerfully as possible."
– Tom Hodgkinson[167]

Whatever job you are doing, seek one that deals with more complex issues, problems, and concepts. That is to say, instead of working in sales for perfumes, cosmetics, books, seek out companies that work with tech, engineering, software, and so on. This does not mean that your profession should change – if you are working in sales, keep doing it, if it pays the bills –, but to actively search for opportunities where you have to deal with topics you do not understand yet. Then you are immersed in such topics as part of your everyday life. Talking from my own experience, in the last few years I have worked for companies dealing with *lost and found-software*[168] – which then inspired me to study business informatics – as well as *building information modeling*[169], *augmented reality*[170], and *software for IVD*[171]. In order to understand why it is worthwhile to deal with complex concepts in today's

---

[167] Hodgkinson, 2005

[168] programmed in C#, utilizing Angular as a front-end framework

[169] BIM is a process of managing digital representations of a buildings or infrastructures

[170] overlay of real-world objects with digital content

[171] IVD: In Vitro Diagnostics is used to perform diagnostic tests on samples taken from the human body

professional world, we need to consider a number of things. First of all, we are aware of the fact that technologies will continue to develop, which means that further specialization in certain technologies and topics will become more important in the future. Therefore, knowledge of such technologies, that is, understanding about the context of use in our society as well as the technological aspects of those technologies, is also necessary. However, it will be believed that a preoccupation with complex concepts is useful only insofar as they benefit us personally in our own careers. This view stems from the ideological standpoint – that one should specialize in a given field instead of becoming a generalist – that is, the current predominant ideology of the ruling class, that one is better off being a small cog in the production process, which can also be easily replaced again. This stems from the motivation of exploitation by the ruling class, that is, becoming an easier to replace part of the production process.

> "The fact that the functions formerly exercised by the totality of the members of the commonwealth become from a certain moment the business of a separate group of people already indicates that there were people who had an interest in carrying on this separation and who profited by it. It is the ruling classes who organize the exclusion of the members of the exploited and productive classes from those functions which would allow them to undo the exploitation imposed upon them."[172]

---

[172] Translated from Mandel (2008): *„Die Tatsache, daß die Funktionen, die früher von der Gesamtheit der Mitglieder des Gemeinwesens ausgeübt wurden, von einem bestimmten Moment an die Angelegenheit einer getrennten Gruppe von Menschen wird, zeigt bereits an, daß es Leute gab, die ein Interesse daran hatten, diese Trennung zu betreiben, und die von ihr profitierten. Es sind die herrschenden Klassen, die den Ausschluß der Mitglieder der ausgebeuteten und produktiven Klassen von jenen Funktionen organisieren, die es ihnen erlauben würden, die Ausbeutung, die ihnen auferlegt wurde, aufzuheben."*

The situation is different if one assumes that one has to educate oneself in many different areas, that is, to build up an understanding in various complex concepts for their own benefit. Here, no consideration is given to whether it helps the ruling class, it does not conform to the prevailing ideology, and thus it is rarely perceived as meaningful. The worker should rather be a specialist, because this way he can be exploited more easily.

"Man, from being "a jack of all trades," has become a specialist. His clothes are made and washed by specialists; his tools are made; his house is built; his food is cooked; his hair cut; and, in some cases the very windows of his house cleaned, by specialists. The details of all these ramifications in the work of modern society would be nearly endless, but this division of labour outside the workshop—not to be confused with the division inside—is all the direct result of the development of trade."[173]

Learning complex concepts is a way to bring back generalism and transform humans into universally useful beings, not just specialists for the exploiting class. It frees you from working only one job to being able to adapt to the complexities, working at jobs that require a multitude of requirements, and therefore become a specialist by providing a unique skillset. The current trend in the job market continues to favor specialists over generalists – a generalist seems to be rather useless nowadays. However, specialists are limited in their scope, and their expertise might become irrelevant as the industry evolves[174]. Learning complex concepts can help you thrive in the ever-changing job market.

---

[173] Starr, 1919

[174] This also means that the job you will do in 20 years from now might be totally different to what you do today.

"Focusing on "using procedures" problems worked well forty years ago when the world was flush with jobs that paid middle-class salaries for procedural tasks, like typing, filing, and working on an assembly line. "Increasingly," according to Duncan, "jobs that pay well require employees to be able to solve unexpected problems, often while working in groups. . . . These shifts in labor force demands have in turn put new and increasingly stringent demands on schools."[175]

Employers are often seeking individuals with in-depth knowledge and skills in specific areas, which allows them to carry out their designated tasks effectively and efficiently. They will try to guilt trip you into focusing on your work at hand instead. Why are you even pursuing something else? You should be doing the work!

"There are some very smart thinkers who believe work is what gives our life meaning because without it we accomplish nothing and make no mark upon history. Without work, we might die and it would be as though we had never lived. Since evolution is fueled by a desire to leave a lasting legacy, this argument can be quite compelling."[176]

However, this trend can be limiting as it restricts our ability to acquire new knowledge and develop diverse skill sets. Such limitations can result in stagnation and hinder the potential for personal and career growth. In contrast, the pursuit of a complex profession that encompasses a broad range of concepts and skills can facilitate the development of a versatile skill set and enhance an individual's adaptability to changing circumstances.

[175] Epstein, 2019
[176] Headlee, 2020

Not everyone will understand your path and that is acceptable too. I distinctly recall a situation where I engaged in a conversation with an individual who derived his income from working in the construction industry. During the conversation, he expressed his disapproval of his daughter's desire to become a lawyer and pursue a career in law, which would require her to undertake formal education. He instead advocated for her to follow in his footsteps and pursue a job in the construction industry that was perceived as secure. This individual's perception was limited by his own experiences and he failed to recognize the potential benefits of pursuing a more complex profession in the future. Furthermore, he had little knowledge of the nature of the legal profession and what it entailed. The aforementioned scenario highlights the notion that individuals may have a limited understanding of the value of pursuing a career that involves learning complex concepts. The conventional wisdom of seeking job security often results in individuals prioritizing the short-term stability of their current job over the long-term potential of their career. This perspective overlooks the dynamic nature of the job market and the continuous evolution of industries, rendering job security an elusive and uncertain concept – the pursuit of a complex profession could prove advantageous as it enables individuals to adapt and excel in an ever-changing landscape. Specialists are experts in their field, with an in-depth knowledge of a specific area. Sometimes the expertise and specializations are even too specific, as the following example shows.

"Highly specialized branches of modern science have become so arcane that many people employed in them are forced to train until their early or mid-thirties in order to join the new priesthood. They may share long apprenticeships, but too often they cannot agree on the best course of action."[177]

Our society is constantly evolving due to a multitude of factors such as technological advancements, economic fluctuations, and changing consumer demands. This leads to more possibilities for people looking for new job opportunities. Predicting these changes and when they occur is impossible – although we have seen ways to understand trends better – and it is therefore essential for individuals to be adaptive. In order to be able to understand latest developments in technology, one must be able to understand complex phenomena, i.e. the multi-faceted concepts and complex ideas that are already established – in existence and in use. The job market is predicted to witness significant transformation in the future, and this is likely to result in the emergence of new and complex professions.

"Eras of great transformation can also spark creativity and innovation."[178]

For example, a report by the *World Economic Forum* that is often cited states, that 65% of children entering primary school today will ultimately end up working in completely new job types that do not yet exist.[179] While *Forbes* states that this statistic seems to be completely made up,[180] we have to make clear that this is an estimate based on a statistic, not a statistic itself. This estimation reflects the information

[177] Smil, 2022
[178] Pop Culture Detective, 2023
[179] World Economic Forum, 2016
[180] Newton, 2018

available at that time. In another Report of 2023 compiled by the *World Economic Forum*, it is estimated that the biggest disruption in the labor market – the macro trends driving business transformation – is the "increased adoption of new and frontier technologies".[181] This dynamic nature of the job market creates both opportunities and challenges for individuals seeking employment, but in order to be able to pursue these opportunities, one must have the relevant skills and knowledge necessary. The pursuit of a complex profession may seem challenging, but it has the potential to provide numerous benefits and opportunities for personal and professional growth.

"There is no royal road to science, and only those who do not dread the fatiguing climb of its steep paths have a chance of gaining its luminous summits."[182]

The current emphasis on job security may be limiting, as it overlooks the dynamic nature of the job market and the continuous evolution of industries. Individuals must seek to develop versatile skill sets and acquire new knowledge, not only to remain relevant and competitive in the ever-changing landscape of the job market, but to have the freedom of choice over your future. Your current employer might think differently about it, might not see it as useful, as we have seen in another scenario.

"We were faced with an explosion of new identities and with complex logics of their articulation that clearly called for a change of ontological terrain."[183]

[181] World Economic Forum, 2023
[182] Marx, 1990

The emerging trends in the labor-market and the predicted transformation in the next twenty years highlight the potential for the emergence of new and complex professions. It is imperative for individuals to remain abreast of these trends and prepare for potential opportunities to excel in their chosen career path.

The perception of education as a means to advance one's career is often limited by the short-term goals of employers and recruiters. This narrow perspective fails to recognize the long-term benefits of continuous education, which can lead to more fulfilling careers, better job security, and improved quality of life. As a case in point, a recruiter once asked me when I would be finished with my studies and what I intended to do afterwards. His inquiry was driven by his interest in knowing whether I would be staying with his company and if my education was immediately relevant to my work. This perspective, while understandable from the employer's point of view, fails to appreciate the broader value of education beyond the immediate demands of the job, because in today's rapidly changing world, education is becoming increasingly important to remain relevant and competitive in the job market – learning is a lifelong process and the skills and knowledge gained from education can be applied to a variety of contexts throughout one's career. As the world becomes more complex and interconnected, the need for individuals with a broader range of skills and knowledge is also increasing. This expertise makes them indispensable to their employers, who rely on them to solve complex problems – but learning something complex and more different than you do today could help making you indispensable in the future.

> "The most remarkable circumstance in our social progress, has been the rapid increase and ascendancy of manufacturing wealth and population. This is the distinguishing feature of society, and to it, I doubt not, may be traced much of the good and evil incidental to our condition – the growth of an opulent commercial, and a numerous and intelligent operative class sudden alternations of prosperity and depression extremes of wealth and destitution – the increase of crime – the spread of education — political excitement —conflicting claims of capital and industry divided and independent opinions on every public question, with many other anomalies peculiar to our existing state. Another result of the transition from agricultural to manufacturing supremacy, has been the creation of not only new interests and new questions of discussion, but also a vast enlargement of the circle of inquirers."[184]

Nevertheless, you will meet some recruiters who value your additional education that falls out of scope of your future employment. They will value your initiative in spending the time and effort necessary to acquire additional skills and knowledge – unlike the recruiter who I was talking about before. This is a rather positive shift in the mindset of recruiters and employers.

> "There are a couple of ways that work can have meaning: [...] work that allows us to develop our talents and interest, to self-actualize, then there is work that grants us autonomy, control over the work itself, and the product of that work [...] there is work that produces something of importance."[185]

---

[184] Wade, 1834
[185] Second Thought, 2023

Whether you will find the right recruiter with such a mindset or not is not necessary. After careful consideration, you should pursue whatever science you see valuable. Who do you want to be in the future?

> "May your dreams not come true, may your hopes not be fulfilled, because they are based on what you know. You should explore possibilities that have never been touched or reached before."
> – Sadhguru

Learning something new and complex is more valuable to yourself than your employer and it will secure your future in the long-run. You could lose your job any time as robots and automation are transforming the business landscape to *Industry 4.0*.[186] However, if anyone asks you at any time why you are still studying, tell them that it is adding more value to your skills while also giving you autonomy over your life and your future. Most people cannot relate why you would be doing all of that effort without thriving for a job opportunity or a financial compensation in the future, but learning for personal interest and enjoyment is just as important as preparing for future employment opportunities, given the constantly evolving landscape of new technologies and industries. That is why, in the next chapters, we will focus on your personal learning endeavor of complex concepts in relation to *activities of leisure*.

> "[…] money can buy anything. Thus our parvenu has everything at his fingertips: beauty, art, knowledge . . . But he could have been a poet, or a painter, or a scientist, someone who would create beauty and

---

[186] Rampersad, 2020

knowledge. [...] In the still-dehumanized life of bourgeois society, money symbolizes this tearing away of man from himself"
– Henri Lefebvre[187]

187 Lefebvre, 2014

An easy gateway to deal with more complex concepts is reading more complex books – books you would normally not read as they are out of scope of what you currently understand. In bookstores, complex books are sometimes very hidden, and online shops only suggest them[188] to you if you have already purchased similar books – which is why you should buy some to have an influence on said algorithm. If you look online you will find many top-reading lists on engineering, math, and science in general. Many YouTube channels give good recommendations as well: Just search for "Best Books on Engineering" and you will find some valuable examples. However, never limit yourself only to the books mentioned in such lists.

Several books have been written on the correct way of reading a book, which I would definitely recommend for the interested learner, e.g. *How to Read a Book* and *How to Read Nonfiction Like a Professor*. In *Beyond Simplicity* I actually want to go into detail why and how to read complex books, which will motivate us to learn even more.

It was shown what advantages it has to deal with complex issues, but how do you incorporate this into your own private life? At first you might think that this is not possible, but we can also deal with complex topics in our free time by reading complex books. In the following chapter I would like to show some of the ways I used to learn about more complex concepts and how you can use them too. I will also show what possible new opportunities this will open in your personal life – which goes far beyond just your career, as discussed previously.

Reading complex *books* is essential because understanding complex concepts helps us not only to understand those concepts, but also to understand how those concepts can be relevant to us. We also experience an expansion of our consciousness and awareness as a result, which further motivates us to deal with even more concepts that we do not yet understand.

When I encourage people to read more, they tend to assume that I am advocating for the consumption of novels. Accordingly, many people reply that they read a lot. Though, what much is to be gained from reading made up stories in novels?[189] I question the value of reading fictional stories when there are other forms of literature that can provide much greater benefits.[190] Although I used to read fantasy and science fiction when I was younger, my reading habits have evolved over time. My first shift occurred when I recognized that I could gain more from reading classic literature and philosophy.

> Studying philosophy improves the student's ability to think clearly, carefully, and logically about a wide variety of topics. It helps to develop the student's ability to assimilate and assess new and unfamiliar ideas and information.[191]

[189] Non-autistic adults prefer fiction over non-fiction according to Chapple et al. (2021).
[190] True for nerds like myself (Mar et al., 2006)
[191] CMICH, 2023

My subsequent shift occurred when I realized that the knowledge I acquired from these sources may not necessarily lead to monetary gain. Although a rather shortsighted view on reading, financial stability was a concern for me at that time, and I began reading self-help literature exclusively. This may seem like a step backward in terms of intellectual substance, but it started a necessary revolution in my writing journey. The writing of Speed Reading Genius startled my interest in delving into philosophical books once again. In recent years, I have found technical and mathematical literature particularly stimulating. This interest was the inspiration for the book you are currently reading. It was mainly the study of business engineering which led me to also reading technical literature in my free time. By all means, even though I was already occupied enough with studying for my degree, reading these books additionally gave me an enormous benefit – the joy of studying complex material.

> "….a good book can teach you about the world and about yourself. You learn more than how to read better; you also learn more about life. You become wiser. Not just more knowledgeable — books that provide nothing but information can produce that result. But wiser, in the sense that you are more deeply aware of the great and enduring truths of human life."[192]

Nobody told me that I should read books about engineering and mathematics and if I had known earlier what positive influence this would have on my intellectual development, I probably would have pursued these topics

---

[192] Adler & Van Doren, 1972

prior. But I was fearing the complexity back then, as my preconceived belief entailed a lack of skill in comprehending particular complex technical literature.

To truly comprehend mathematical books and engineering literature, one must engage in active study and problem-solving of these texts. The material is not readily accessible and requires a deep understanding of the interrelationships between the complex concepts in order to progress to more advanced topics. One of the benefits of studying such literature is the development of problem-solving skills[193] – by solving mathematical problems you are then able to apply that skill to other areas of life.

> "Finding a way to divide a problem into steps or phases can make the problem much easier. If you can divide a problem into two pieces, you might think that each piece would be half as difficult to solve as the original whole, but usually, it's even easier than that."[194]

When a book requires significant intellectual effort, it does not necessarily imply a lack of intelligence. Never feel bad for taking longer to comprehend the material. Through the act of decomposing the problem into smaller problems, clarifying and solving each and everyone of them, to finally come to a solution, you are gaining a valuable learning experience[195] that fosters perseverance and resilience in the face of challenges and setbacks. It will likewise encourage you to learn even more because you now trust yourself that you can do it. Truly, the ability to read and comprehend

---

[193] Căprioară, 2015
[194] Spraul, 2012
[195] Dostál, 2015

mathematical literature is a highly valuable skill that can be applied to a wide variety of branches.

> "Whether chemists, physicists, or political scientists, the most successful problem solvers spend mental energy figuring out what type of problem they are facing before matching a strategy to it, rather than jumping in with memorized procedures."[196]

The question of how to read correctly goes beyond the simple act of consuming content from front to back. Reading can be an active and strategic process that involves critical thinking and analytical skills. The effectiveness of reading largely depends on the reader's approach and the context of the text. While reading nonfiction books, it is essential to approach the text with a specific purpose and an open mind to different perspectives. It involves active engagement with the text, especially in relevance to the purpose of reading it. This purpose may be to gain new insights into a particular subject, understand a particular complex concept, or gather information to support an argument.

> "Reading exposes you to lives you'd never known before, experiences you'd never imagined, and modes of thinking far different from your own. All of this builds both your empathy for others and your understanding of how the world works beyond yourself."[197]

In this context, reading books that align with one's *vision* is vital in broadening one's perspective. I personally own several such books that my *vision* would read but which I haven't yet read myself. I want to be surrounded by books that

---

[196] Epstein, 2019
[197] Kwik, 2020

I don't yet understand because they motivate me to learn even more to eventually understand them. That means, of course, I have to read many more books to get there, to understand those books. In this case, it is a special kind of category of books in my library that I call *Książki* - which is the Polish word for books. I always refer to them as *motivation*.

It is also important to not to be impressed by apps that summarize complex books that we otherwise would not have the time to read. This does not help us as we need to engage with such books in depth. Summarizing a book into a short format results in simplifications and generalizations – something this book goes highly against. I want you to understand that condensing books into bite-sized summaries sacrifices the depth and nuance of the original material, which can be crucial for a comprehensive understanding of complex subjects. Supporting materials, such as evidence, research, and examples to bolster arguments, are often excluded, left out, and the reader is left behind. Reading a book in its entirety can create a more immersive and personal experience, allowing readers to connect with the author's ideas and experiences. Summary apps, on the other hand, may fail to foster the same level of personal connection and emotional resonance. To truly acquire knowledge, we must resist the allure of shortcuts and embrace the intellectual challenge of engaging with complex books.

"The mode of production in material life determines the general character of the social, political, and spiritual processes of life. It is not the consciousness of men that determines their existence, but, on the contrary, their social existence determines their consciousness."[198]

198 Marx, 1859

If our consciousness is influenced by our environment, then surely we could change our environment so that our consciousness changes as a result – that's why it's important to fill our own library with intellectual books that we want to understand one day, even if we don't understand them today. What would the person you want to be – your *vision* – read? By putting those very books on one's shelf, one promotes the consciousness of being that very person – even if you might never read those books. A shelf of unread books is not essential if it only exists to satisfy our ego. However, we should use them as motivational books to change our consciousness by surrounding ourselves with those books that our *vision* would read. We form our environment according to our own ideas - like a sculptor who forms his sculptures – and this way we ultimately *form ourselves*. In the end, it is not about how many books one reads, but rather what one reads - and what one *makes* of it.

"Every one of us possesses a tremendous amount of knowledge about the world. Some of our basic skills are determined by our genes, such as how to eat or how to recoil from pain. But most of what we know about the world is learned."[199]

Higher education offers much more than just learning a specific thing like programming languages or algorithms. If it is in any way possible for you, you should consider enrolling in such a program – science, engineering, math. Of course, it also provides a prerequisite for an interesting job, but that is not the main motivation of this book. But maybe you do not want to commit to a degree program, or perhaps it is not easy for you to pursue a degree where you live due to higher financial barriers. In that case, distance learning platforms, online course platforms, and similar options provide an alternative, such as *Coursera, Udemy, EdX, FutureLearn, MIT OpenCourseWare, Fernuni Hagen,* and *University of the People.* Unlike in the United States, we are fortunate here in Austria that tuition fees are very low or non-existent. At universities, they have been abolished for those who study and finish their degree within the minimum duration of the program. Then there are universities of applied sciences[200] where semester fees are charged, but they are by no means as high as in the United

---

[199] Hawkins, 2021

[200] Higher education in *Fachhochschulen.* These are not considered the same as universities, but due to the limitation of language there is no more suitable word available here.

States. Nevertheless, many people still choose not to pursue a degree and some students start studying but do not complete their degree. This chapter is dedicated to those readers who would have the privilege of choosing to pursue a degree due to the opportunities available to them, but still choose not to do so.

I realized the numerous advantages of a university education rather late, especially since I resumed studying after many years of hiatus. Motivated not by career redirection, but rather by a fervent desire to delve deeper into neglected domains, I soon discovered that my educational journey encompassed more than the mere acquisition of subject matter expertise. During this process I realized that I was learning much more than just the content of the course – I was also gaining self-confidence, becoming more self-assured, and willingly taking on greater *challenges*. Within the course of a higher education in engineering and sciences, one becomes intimately acquainted with a multitude of intricate and multifaceted topics – such an extent and variety may not be as apparent in the subsequent journey of your professional career. It is within this transition that the discerning eye may perceive a tempered manifestation of the once profound mosaic of knowledge, as the contours of professional specialization begin to carve a narrower path. The indelible mark of holistic comprehension and the cognitive dexterity fostered by such a comprehensive education remain invaluable assets in navigating the labyrinthine complexities of the engineering realm. Truth to be told, never in your life will you ever have this multitude of complex topics interrelated to each other, unless of course your pursue an academic career of a researcher or scientist.

Choosing an appropriate study program at any depends on a variety of factors which need to be taken into consideration. First of all, you need to focus on something that truly excites you. However, this can be a biased view as you do not yet know what you are obsessed with if you have never been exposed to it. My fascination for programming only appeared after my experience working in a software company. It became crucial for me to unravel the inner workings of software programs and comprehend their underlying mechanisms. It was during my work as a technical writer at this very company that I engaged in a conversation with an individual who was, at that time, busy in crafting a master's thesis on *Topology*. This conversation sparked a first interest in the topic, propelling me towards delving deeper into this topic. However, it was much later that I began to seriously study this subject matter[201]. Now, I was fortunate enough to come across these concepts in such a manner. But what should someone study if they are entirely unfamiliar with a concept or never exposed to it? How can they begin to grasp its allure or significance? A first clue can be found in the books that one has on their shelves at home[202]. When you are not exposed to new concepts at your work, it could also be a fascination with certain technologies, such as *Wi-Fi*, *microwaves*, *rockets*, and so on, that can inspire us to study something in greater detail.

---

[201] I consulted several books, watched videos, and even applied for a study program on Mathematics at *Fernuni Hagen*, to be seriously engaged in the topic of Mathematics although I know would have had a class on topology much later in the course. When I am talking about "seriously" studying a subject matter, I mean engaging with it deeply and putting your whole heart into it.

[202] If you do not own any books on *science*, *math*, *technology*, or *engineering*, you should change that immediately.

The ability to grasp complex concepts is intertwined with the capacity to discern the technologies that will hold greater relevance in the future. Because only when we understand the extent of a technology and its significance can we comprehend and appreciate what that means for the future of humanity as a whole. If someone has no idea about a new technology, it might seem incredible simply because it is new and unique. Only through an appreciation of a technology can we ascertain its future significance. However, this understanding is crucial in deciding what to study because we naturally want to pursue something that will remain relevant in the future. How can we develop this foresight?

> "Every successful program you write is more than a solution to a current problem; it's a potential source of analogies to solve future problems."[203]

The ability to perceive something that does not yet exist lies in the capacity to extract additional insights from existing information, rather than interpreting more into it. It is the understanding of a subject that surpasses our immediate scope of application, whether it is due to its complexity, making it difficult to find practical uses in our everyday lives, or its intricacy, preventing us from immediately comprehending its potential applications even outside our daily lives. This is a good indication to delve deeper into that matter. If, for example, this applies to mathematics, it does not mean that one should study mathematics per se, but rather choose a field of study that incorporates a substantial amount of mathematics. In other words, what we should focus on is

[203] Spraul, 2012

mathematics as a driving and essential force of a particular discipline. Mathematics, being nothing other than formal logic, thus provides an important foundation for understanding this subject.

> "Young man, in mathematics
> you don't understand things.
> You just get used to them."
> – John Von Neumann

A field of study without mathematics, such as sociology, philosophy, history, and so on, may teach us about those subjects, but it will hardly provide us with the necessary knowledge of logic required for comprehending complex concepts. Therefore, for the understanding of complex concepts, which is our main focus here, a study program whose fundamental component is mathematics is essentially indispensable. There are many ways on how to improve abstract thinking and studying mathematics is one of them. My personal experience has led me to recognize the inherent beauty and elegance of mathematics, and its unique ability to provide insight into the workings of the natural world. As such, I have developed a deep appreciation for the discipline and its many fascinating applications. By studying mathematics, I improved my ability to think logically and systematically. This is especially valuable in today's fast-paced and complex world, where the ability to think critically and creatively is highly prized. It is also something that is beneficial for a wide range of other sciences. So if you are studying anything that includes mathematics, that is formal logic, you will see its immediate benefit when you study other sciences. This, of course, should be the main motivation to study any particular scientific topic – gaining a better overall understanding of all the sciences.

> "One who understands nothing but chemistry, does not understand chemistry."[204] – Georg Christoph Lichtenberg

On the one hand, one wants to precisely define and divide the sciences – they should not be mixed. And yet, within these sciences, not only are distinct fields created, but they are also interpreted differently. It also depends on the historical context in which a book about a certain topic is written.

---

[204] „Wer nichts als Chemie versteht, versteht auch die nicht recht."

In today's day and age, it is frequently contemplated to write for a specific target audience, one that can extract particular value from the work that is being created. This entails crafting *products*[205] with the potential to make money from selling them. The driving force behind this motivation is to *produce* a work that may serve as an exceptionally useful asset to society – to create *value* that people will *pay* for accordingly. Nevertheless, the reader will find that the general public will likely show little to no interest in the content of one's writing. Writing is a personal endeavor, meant to be enjoyed for its own sake.

"Think of yourself not just as a taker of notes, but as a giver of notes – you are giving your future self the gift of knowledge that is easy to find and understand."[206]

Creating *products* in the sense of making something to sell for, is not the aim of this chapter. However, one can take pleasure in the creation process and at least *imagine* that there will be people who will engage with the works – that is books, lectures, and so on –, read them, and hopefully reflect upon them.

---

[205] *Creations* of your *productive* and *creative* mind.
[206] Forte, 2022

It is through the creation of *output* that a genius has unfolded their true capacity of knowledge and understanding – ergo, the significance of a genius high intelligence and their brilliant mind resides in their capacity to generate *output*[207] that transcends the boundaries of conventional thought. This *output*, born from the ingenuity and transformative capacity of the genius, transcends mere functionality to embody utility of the highest order. *Output* encompasses all productive endeavors such as drawing, writing, constructing, and the like – activities that manifest thoughts onto a medium, enabling the measurement of one's productivity. It encapsulates the diverse array of creative manifestations and transcends the realm of the mundane. It is through the synthesis of intellectual endeavors and scientific research, that the genius is able to craft their *output*, ultimately fostering a legacy that transcends their mortal existence. And through the creation of such *output*, the genius is able to assimilate existing knowledge, concepts, and resources into an intellectual novelty. At the end it could evoke awe, reverence, and contemplation among those fortunate enough to engage with it. However, much more significant than that is the transformation that one undergoes oneself through the creation of such a work.

> "The totality of objects and human products taken together form an integral part of human reality. On this level, objects are not simply means or implements; by producing them, men are working to create the human; they think they are moulding an object, a series of objects – and it is man himself they are creating."[208]

---

[207] The culmination of such cognitive prowess is exemplified in a tangible *output* – a book or something similar, which encapsulates the essence of a genius.

[208] Lefebvre, 2014

I would like to emphasize the significant aspect of learning through the creation of such *output*. This is something that I was not aware of until I experienced it myself. No one told me to write a book. I began by emulating and imitating geniuses, endeavoring to write more and more about the things I wanted to learn. It turned out that through this method – namely, writing a book – I learned, realized, and understood much more than if I had simply studied it passively.

> "If we simply put our knowledge and opinions on paper, we may learn the errors and deficiencies before others have a chance to see them. In order to describe our thoughts, opinions, and knowledge on paper, one must think each step through to its logical conclusion. In this process, the writer often can become aware of errors which he had not recognized before."[209]

It is primarily through the endeavor of wanting to write a book that one must engage in thorough research and understanding of the subject matter. This involves delving into complex topics in detail rather than merely scratching the surface. Moreover, when you set yourself a goal of having written a book with a minimum of a hundred, two hundred, or even more pages, you must write more then, meaning you will also have to learn more. Writing a book on a scientific topic is one thing but we also want to publish a new finding or insight, and I think that is a great motivation to learn more. This necessitates expanding on ideas and showcasing connections to other subjects. As a result, you will learn much more and retain it more effectively as well.

[209] Boucher, 1972

"The point of writing is to discover something."
– Shaun Levin

What is worth writing down and what is not? Finding time to *write* means finding time to *think*. Writing is a meditative practice and is often referred to as writing from the soul. Meditation helps us focus on a single thing, and with writing, thoughts transfer onto paper. Stress dissipates, leaving behind brilliant ideas that can be further developed., i.e. we can reflect upon these ideas. Without reflection, one has nothing to write about, which sounds obvious the more you think about. Therefore, more emphasis should be placed on reflection, that is the thinking about something, rather than writing. However, this is only partially true. By writing more, we engage in thinking – *on paper*. *Output* is measurable. It makes *thinking* measurable too. The more *output* we create, the more we have contemplated about a subject. To generate more *output*, we do not solely focus on the act of creation but on the subject matter itself – we *think* about it – and learn more in the process. *Write* as if you were explaining the topic to others. Use simple words and illustrate the connections to other areas, or within the same area, depending on the complexity of the concepts you want to present. By putting it down on paper and explaining it to your target audience, you can determine if your explanation is coherent. Again, you will understand and remember a lot in the process. The question remains: should you publish your book[210] or let it rot on your hard drive?

---

[210] You could do this via KDP, IngramSpark, Lulu, or other *print-on-demand* publishing services.

"Leibniz published just one philosophical book, the Theodicy, and a handful of articles in learned journals; otherwise much of his work did not see the light of day until long after his death [...] Because of Leibniz's reluctance to publish, his early readers had access to only a fraction of his total output and therefore acquired a one-sided and misleading impression of his achievement."[211]

Perhaps the format of a book is not suitable for everyone; in some cases, a *treatise* would suffice. This can be a scientific paper or work, which, however, does not have to reach the dimensions of a book – that is to say, I would consider a treatise a very simple yet effective discourse on a specific *scientific* topic. However, we have seen from the past, that this is not necessarily the case. Many treatises appear to be very sophisticated, – such as monographs – bringing forth the most recent insights and findings in a scientific field or defining it in its completeness.

"During his lifetime, Leibniz produced treatises of great value on the widest possible range of subjects."[212]

The question remains as to how meaningful the publication of a book really is and whether it contributes directly to the information overload without actually disclosing truly relevant scientific insights. This question is certainly justified. However, if one assumes that the true purpose of this work – namely, the creation of a treatise on a complex field of study – is rooted in learning more about a subject, then the publication of a work is merely a byproduct

[211] Jolley, 2008
[212] Jolley, 2008

of this endeavor. So it's not so much about how others might perceive it, or how it is perceived by the general public, but rather about how beneficial it has been for you, the creator of such work. It is often times the case that finding your niche is an issue when becoming a successful author. Honestly, I do not believe in such limitations and if you want to write in engineering or philosophy, then by all means write – as did Gottfried Wilhelm Leibniz:

> "From philosophy and mathematics narrowly construed [the works of Leibniz] extended across the encyclopedia of the sciences and beyond: to astronomy, physics, chemistry, and geology; to botany, psychology, medicine, and natural history; to jurisprudence, ethics, and political philosophy; to history and antiquities, German, European and Chinese languages; to linguistics, etymology, philology, and poetry; to theology both natural and revealed; and beyond contemplative pursuits altogether to a wide range of practical affairs: from legal reform to the reunification of the churches, from diplomacy and practical politics to institutional reform, technological improvement, and the organisation of scientific societies, libraries, and the book trade."[213]

If you believe writing a scientific book or paper is too much of a hassle, start by just writing down your thoughts and questions on a particular subject. These form a great outline for a book. Many initial ideas originated on the fringes of science because the individuals involved posed questions that personally affected them at the time and they attempted to answer them too.

> "All these elementary movements of the working class have largely been led by the workers themselves, i.e., by autodidacts who often formulated naive thoughts about historical, economic, and social

---

[213] Antognazza, 2011

questions that actually require scientific studies to be thoroughly addressed. Therefore, these movements develop, in a way, on the fringes of the scientific progress of the 17th and 18th centuries."[214]

Writing is a way to record your thoughts, communicate your insights and spread your ideas. Once you understand that, writing becomes easy. All you have to do is to write down what you are thinking for a longer period of time, expand your thoughts and express them in more detail later on. You organize what you are thinking and a book is a complete collection of your ideas – a product of your creative mind compressed in a readable and understandable format which can be replicated, shared, read, and so on. You do not have to be a great writer in order to be successful at employing writing as a learning method. Some questions to start with writing are:

- What is this about?
- How does it work?
- Why is it popular?
- What is missing?
- How could it be improved?
- Why has nobody thought of this before?

You might come up with even more question which will lead you to undiscovered territory. From that point on, writing a book is an easy task. Creating more works in different

[214] Mandel, 2008 - *"Alle diese elementaren Bewegungen der Arbeiterklasse sind weitgehend von den Arbeitern selbst geleitet worden, d.h. von Autodidakten, die häufig naive Gedanken über historische, ökonomische und soziale Fragen formulierten, die eigentlich wissenschaftliche Studien erfordern, um gründlich behandelt zu werden. Diese Bewegungen entwickeln sich daher gewissermaßen am Rande des wissenschaftlichen Fortschritts des 17. und 18. Jahrhunderts."*

formats greatly increases your chance of achieving genius status. Also, having different works and formats allows you to appeal to different audiences, that is to say, they will consume what you have created at some point more easily if you provide your *output* in different *formats*[215]. Write about topics that interest you, no matter what they are. This is the only way to ensure that the *products* you create have lasting value. Productivity means constantly developing new *products* and publishing them. You will learn to think more critically – and this in turn will help you build your own understanding of the interconnectedness of complex concepts[216].

"Either write something worth reading
or do something worth writing."
– Benjamin Franklin

---

[215] On a side note: Not everyone will like all your works, no matter how much time you invest in them; however, writing books on different topics – which you connect with their original ones – can help you reach a larger number of readers and drastically increase your visibility.

[216] This highlights the importance of exploring existing concepts beyond your understanding and improving upon them.

Applications:
My Personal Challenges

For the final chapter I would like to focus on giving a few personal examples of *challenges* that I have defined for myself and which I am eager to solve. Additionally I would like to present my learning philosophy a little bit more, that is my own attitude towards learning in general.

In the past I would have said that "writing at least 150 books[217]" could be such a challenge, but after defining it more properly in the last chapter, I came to the conclusion that writing books is not a challenge by itself, but rather a means of facing my *vision* of *becoming a genius*. Nevertheless I have identified certain *challenges* as a means of solving problems which I want to outline below:

- Write a highly intellectual book on a new science of genius[218] to solve the question what a genius is and how to become one once and for all.
- Write a book on why it is important to study complex topics[219] to solve the problem that people do not want to deal with complex information or give up too easily.

---

[217] In the book *Speed Reading Genius* I have declared that *books* are the main reason why we can identify a genius today. One can say of him or herself that they are the most intelligent being ever lived on this planet but you would need some kind of proof of that – and *output* gives this exact proof.

[218] I have done this in German and an English translation will soon follow.

[219] You are reading it right now.

- Write a book on Slavoj Žižek and his thinking to analyze and solve the question what exactly the philosophy of Žižek is.
- Write a book about learning methods for people on the autism spectrum to help overcome learning difficulties.
- Write a book on extinct indigenous languages to make these languages more popular again and make people more aware about *Sprachensterben*[220].

Please take note that I will probably not become rich by solving these challenges and that this is certainly not my goal or motivation in this endeavor. By presenting you this list I want to motivate you to create your own challenges and create your own *output*. I am not afraid that you would steal my list of ideas as ideas are relatively worthless if they are not implemented. This large pool of ideas allows me to have the freedom to choose one or work on one whenever I want to be relatively free in shaping the list and adding new book ideas. There are just so many topics I want to deal with but I do not have to worry to put everything inside one book alone as I could always create and publish a new book.

Looking forward to something, cultivating an open mind, and so on... All this is important as part of the *philosophy of learning*, which includes one's definition of learning and other connecting points. Likewise, it includes the why, that is, the motivation for learning on an objective level. Next comes the psychology of learning, which deals with the subjective motivation of the individual and explains

---

[220] Languages are dying.

how to maintain it in the long term. Finally, there would be the learning methods. If we were to summarize it, it would look like this:

- *Philosophy of learning:* basic attitude towards learning.
- *Psychology of learning:* own motivation to learn something or ways to increase this motivation.
- *Learning methods:* methods and techniques to learn the subject.

You could say that a learning plan is one of the learning methods - because organizing the learning material would be a learning method. But which learning methods to take is a question of learning philosophy and learning psychology. For example, I believe that books are better than videos (learning philosophy) and that I learn more from writing than just from reading (learning psychology). A clear delineation is not always possible on the part of the subject matter.

In this chapter I focus more on *learning philosophy* and *learning psychology*. We have looked at *learning methods* for complex concepts in another part of this book, although they might not be as important as the other two, that is to say, if you know why you are learning (*learning philosophy*) and you are motivated to learn (*learning psychology*), then you will find a way (*learning method*) to do it. Below are some examples of ideas of my own *learning philosophy*. To succeed on your learning journey, you should definitely create your own:

1. The first rule to success is: Start.
2. The second rule is: Don't stop.
3. Learning is never over. It is a lifelong journey.

4. It is better to start right away - even if it is imperfect - than to try to do it perfectly from the beginning.

5. Reading words you don't understand is relaxing. It gives a sense of humility. To finally understand them is encouraging. One will encounter both feelings.

6. You can read complex books for fun. Get used to not understanding things right away. Take pleasure in not always having to understand everything right away.

7. If you don't understand something the first time, it doesn't mean you are stupid. It means there is an opportunity to learn something new.

8. Never compare yourself to what others are doing. Stop focusing on others.

9. The best time to start is right now. Not later. Don't put off your project.

10. If you feel motivated to learn something, it means you have an inner desire to figure something out and solve a mystery.

11. Every learning challenge brings you one step closer to your goal.

12. The difficult things are hard only at the beginning. Once you understand them, they becomes easy.

13. Time is limited. Energy is limited. The capacity of your brain is unlimited. What does this mean? It means that you must manage what is limited.

14. Teach yourself. Don't be dependent on others. Become an autodidact.

Conclusion

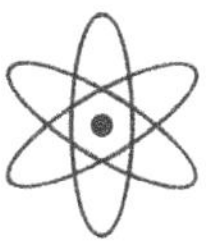

Thank You
for Reading This Book

"If you want to teach people a new way of thinking, don't
bother trying to teach them. Instead, give them a tool, the use
of which will lead to new ways of thinking."
Richard Buckminster Fuller

Conclusion

It is my hope that this book has inspired and equipped you with the necessary tools to embrace complex topics with more confidence and enthusiasm. By applying the strategies and techniques outlined in this book, you can empower yourself to tackle even the most daunting subjects.

In this book, we have explored the challenges individuals face in grasping intricate subjects, particularly in the realm of science. We began by acknowledging the declining interest in science and the implications it has for our society. We then delved into the importance of cultivating a focus on scientific topics and their broader significance. We examined intellectual challenges that hinder our ability to comprehend complex topics and discussed effective strategies to overcome them. We explored the power of abstract thinking and the role it plays in unraveling intricate concepts. We also emphasized the significance of learning fundamental principles and systems as a foundation for grasping complex subjects. We discussed how a deep understanding of complex subjects can lead to enhanced problem-solving abilities, innovative thinking, and improved decision-making skills. We also highlighted the value of incorporating complex topics into our reading habits to foster intellectual growth and broaden our perspectives.

As you continue your intellectual learning journey, remember that you have the power to shape the future through your understanding of science and its intricate subjects. By engaging with these topics and by sharing your knowledge through writing and the creation of *output*, you can contribute to the collective progress of humanity. Now, armed with the insights gained from this book, it is time to embark on your own personal quest for knowledge. Embrace the challenges, celebrate the victories, and never cease in your pursuit of understanding complex topics.

The pursuit of knowledge is not a linear path but rather a continuous cycle of exploration, application, reflection, and refinement. It requires curiosity, open-mindedness, and an unwavering commitment to intellectual growth. The world awaits your discoveries – so innovate and contribute! Remember, learning and understanding complex topics is not a solitary endeavor. Engage with communities of learners, seek guidance from experts, and participate in meaningful discussions. Together, we can create a society that values and embraces the wonders of science, nurturing a future filled with boundless possibilities. May your journey be filled with intellectual growth, discovery, and a deep appreciation for the intricacies of the world around us. Let's embrace the challenge of learning and understanding complex topics by going *beyond simplicity*.

Remember that learning complex concepts may be
challenging,
but the rewards will be *immeasurable*.

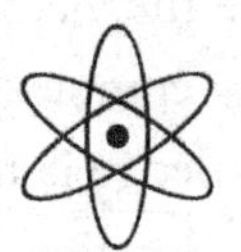

Bibliography

[1] Anderson (1976): *Perry Anderson*, Considerations on Western Marxism, Verso, 1976

[2] Arias et al. (2004): *Pedro Arias, Cora Dankers, Pascal Liu, Paul Pilkauskas*, The World Banana Economy 1985–2002, FAO Commodity Studies, 2004

[3] Asshaari et al. (2012): *Izamarlina Asshaari, Norngainy Mohd Tawil, Haliza Othman, Nur Arzilah Ismail, Zulkifli Mohd Nopiah, Azami Zaharim*, The Importance of Mathematical Pre-University in First Year Engineering Students, Procedia - Social and Behavioral Sciences, Volume 60, 17 October 2012, Pages 372-377

[4] Adler & Van Doren (1972): *Mortimer J. Adler & Charles Van Doren*, How to Read a Book: The Classic Guide to Intelligent Reading, Touchstone, 1972

[5] Adler (2009): *Paul Adler (ed.)*, Marx and Organization Studies Today, The Oxford Handbook of Sociology and Organization Studies: Classical Foundations, March 2009

[6] Aiton (1985): *E J Aiton*, Leibniz – A Biography, Adam Hilger Ltd, Bristol and Boston, 1985

[7] Antognazza (2011): *Maria Rosa Antognazza*, Leibniz: An Intellectual Biography, Cambridge University Press, December 2011

[8] Barabási (2023): Albert-László Barabási, Why Einstein is a "peerless genius" and Hawking is an "ordinary genius", Big Think, https://www.youtube.com/watch?v=XsBcdfKfy9o *Accessed on 10. June 2023*

[9] Berkeley (2023): How Thinking About the Future Makes Life More Meaningful, https://greatergood.berkeley.edu/article/item/how_thinking_about_the_future_makes_life_more_meaningful, *Accessed on 5. May 2023*

[10] Bodemanm (1889) *Eduard Bodemann*, Der Briefwechsel des Goutfried Wilhelm Leibniz der Königlichen öffentlichen

Bibliothek zu Hannover. Hanover, 1889; repr. Hildesheim: Olms, 1966

[11] Borgstroem (1921): *Arthur Travers-Borgstroem*, Mutualism: A Synthesis, Macmillan and Co, London, 1921

[12] Botkin et al. (1979): *James Botkin, Mahdi Elmandjra, Mircea Malitza*, No Limits to Learning: Bridging the Human Gap, Oxford, UK, Pergamon, 1979

[13] Boucher (1972): *Carl O. Boucher*, Writing as a means for learning, The Journal of Prosthetic Dentistry, Volume 27, Issue 2, February 1972, Pages 229-234

[14] Burke (2021): *Peter Burke*, The Polymath: A Cultural History from Leonardo da Vinci to Susan Sontag, Yale University Press, 2021

[15] Căprioară (2015): *Daniela Căprioară*, Problem Solving - Purpose and Means of Learning Mathematics in School, Procedia - Social and Behavioral Sciences, Volume 191, 2 June 2015, Pages 1859-1864

[16] CBL (2023): https://www.challengebasedlearning.org *Accessed 15. May 2023*

[17] Chang & Tsai (2022): *Hao-Tien Chang, Fu-Ching Tsai*, A Systematic Review of Internet Public Opinion Manipulation, Procedia Computer Science, Volume 207, 2022, Pages 3159-3166

[18] Chapple et al. (2021): *Melissa Chapple, Sophie Williams, Josie Billington, Philip Davis, Rhiannon Corcoran*, An analysis of the reading habits of autistic adults compared to neurotypical adults and implications for future interventions, Research in Developmental Disabilities, Volume 115, August 2021

[19] Clark (2021): *Dorie Clark*, The Long Game, Harvard Business Review Press, September 2021

[20] Clarke (1964): *Arthur C. Clarke*, Profiles of the Future: An Inquiry into the Limits of the Possible, Bantam Books, 1964

[21] CMICH (2023): College of Liberal Arts & Social Sciences, Why Study Philosophy?, https://www.cmich.edu/academics/colleges/liberal-arts-social-sciences/departments/philosophy-anthropology-religion/philosophy/why-study-philosophy *Accessed on 22. May 2023*

[22] Collison & Nielsen (2018): *Patrick Collison, Michael Nielsen*, Science Is Getting Less Bang for Its Buck, The Atlantic, November 16, 2018.

[23] Contessa (2022): *Gabriele Contessa*, The Reasons for Science Skepticism Can be Complex, https://theconversation.com/the-

reasons-for-science-skepticism-can-be-complex-and-founded-on-real-concerns-171000, January 4, 2022

[24] Dean (2019): *Jodi Dean*, Comrade: An Essay on Political Belonging, Verso, 2019

[25] Ditta et al. (2020): *Annie S. Ditta, Carla M. Strickland-Hughes, Cecilia Cheung, Rachel Wu*, Exposure to information increases motivation to learn more, Learning and Motivation, Volume 72, November 2020

[26] Dostál (2015): *Jiří Dostál*, Theory of Problem Solving, Procedia - Social and Behavioral Sciences, Volume 174, 12 February 2015, Pages 2798-2805

[27] Eckermann (1885): *Johann Peter Eckermann*, Gespräche mit Goethe am Ende seines Lebens, Brockhaus, Leipzig, 1885

[28] Epstein (2019): *David L. Epstein*, Range: Why Generalists Triumph in a Specialized World, Riverhead Books, 2019

[29] Erard (2012): *Michael Erard*, Babel No More: The Search for the World's Most Extraordinary Language Learners, Free Press, New York.

[30] Escolà-Gascón et al. (2021): *Álex Escolà-Gascón, Neil Dagnall, Josep Gallifa*, Critical thinking predicts reductions in Spanish physicians' stress levels and promotes fake news detection, Thinking Skills and Creativity, Volume 42, December 2021

[31] Fisher (2009): *Mark Fisher*, Capitalist Realism: Is There No Alternative?, Zero Books, 2009

[32] Forte (2022): *Tiago Forte*, Building a Second Brain, Profile Books, 2022

[33] Geisel (1978): *Theodor Seuss Geisel, Dr. Seuss*, I Can Read With my Eyes Shut, Random House Books for Young Readers; First Edition, 1978

[34] Greene (2013): *Robert Greene*, Mastery, Penguin Books, 2013

[35] Gupta et al. (2022): *Ashish Gupta, Han Li, Alireza Farnoush, Wenting Jiang*, Understanding patterns of COVID infodemic: A systematic and pragmatic approach to curb fake news, Journal of Business Research, Volume 140, February 2022, Pages 670-683

[36] Haleem et al. (2022): *Abid Haleem, Mohd Javaid, Mohd Asim Qadri, Rajiv Suman*, Understanding the role of digital technologies in education: A review, Sustainable Operations and Computers, Volume 3, 2022, Pages 275-285

[37] Hamming (2020): *Richard W. Hamming*, The Art of Doing Science and Engineering, Stripe Press, 2020

[38] Haverbeke (2018): *Marijn Haverbeke*, Eloquent JavaScript: A Modern Introduction to Programming, No Starch Press; 3rd Edition, 2018

[39] Hawkins (2021): *Jeff Hawkins*, A Thousand Brains: A New Theory of Intelligence, Basic Books, March 2021

[40] Headlee (2020): *Celeste Headlee*, Do Nothing: How to Break Away from Overworking, Overdoing, and Underliving, Harmony Books, 2020

[41] Hegel (1977): *G.W.F. Hegel*, Phenomenology of Spirit, Oxford University Press, 1977

[42] Herman (2019): *Todd Herman*, The Alter Ego Effect: The Power of Secret Identities to Transform Your Life, Harper Business, 5. February 2019

[43] Ho & Sculli (1997): *J.K.K. Ho, D. Sculli*, The scientific approach to problem solving and decision support systems, International Journal of Production Economics, Volume 48, Issue 3, 14 February 1997, Pages 249-257

[44] Hodgkinson (2005): *Tom Hodgkinson*, How to be Idle, Penguin Books, 2005

[45] Hogancamp et al. (2022): *Matthew Hogancamp, David E. V. Rose, Paul Wedrich*, A Kirby color for Khovanov homology, arXiv:2210.05640, 25 Oct 2022

[46] Jain et al. (2011) *A. Jain, S. Soner, A. Gadwal*, Reverse engineering: Journey from code to design, 3rd International Conference on Electronics Computer Technology, Kanyakumari, India, 2011, pp. 102-106

[47] Jolley (2008): *Nicholas Jolley*, Cambridge Companion to Leibniz, Cambridge University Press, 12. January 2008

[48] Kaczynski (2018): *Theodore John Kaczynski*, Unabomber, Netflix, 2018

[49] Kafatos & Yang (2016): *Menas C. Kafatos, Keun-Hang Yang*, The quantum universe: philosophical foundations and oriental medicine, Integrative Medicine Research, Volume 5, Issue 4, December 2016, Pages 237-243

[50] Kakeru (2019): Creature Girls: A Field Journal in Another World, Vol. 1: A Hands-On Field Journal in Another World, Ghost Ship, 2019

[51] Keller et al. (1991): *Janet Dixon Keller, F.K. Lehman, U Chit Hlaing*, Complex concepts, Cognitive Science, Volume 15, Issue 2, April–June 1991, Pages 271-291

[52] Krauss (2007): Lawrence M. Krauss, Fear of Physics, Basic Books, 2007

[53] Kwik (2020): *Jim Kwik*, Limitless: Upgrade Your Brain, Learn Anything Faster, and Unlock Your Exceptional Life, Hay House, 2020

[54] Laclau (2014): *Ernesto Laclau*, The Rhetorical Foundations of Society, Verso, 2014

[55] Lefebvre (2014): *Henri Lefebvre*, A Critique of Everyday Life, Verso, 2014

[56] Legaki et al. (2020): *Nikoletta-Zampeta Legaki, Nannan Xi, Juho Hamari, Kostas Karpouzis, Vassilios Assimakopoulos*, The effect of challenge-based gamification on learning: An experiment in the context of statistics education, International Journal of Human-Computer Studies, Volume 144, December 2020

[57] Loewenstein (1994): *George Loewenstein*, The Psychology of Curiosity: A Review and Reintrepretation, Psychological Bulletin. 1994, Vol. 116 No.1 p. 75–98

[58] López-Fernández et al. (2020): *D. López-Fernández, P. Salgado Sánchez, J. Fernández, I. Tinao, V. Lapuerta*, Challenge-Based Learning in Aerospace Engineering Education: The ESA Concurrent Engineering Challenge at the Technical University of Madrid, Acta Astronautica, Volume 171, June 2020, Pages 369-377

[59] Lukács (2009): *Georg Lukács*, Lenin: A Study on the Unity of His Thought, Verso, 2009

[60] Maes et al. (2011): *Katrien Maes, Koenraad Debackere, Paul van Dun*, Universities, Research and the "Innovation Union", Procedia - Social and Behavioral Sciences, Volume 13, 2011, Pages 101-116

[61] Maltz (1978): *Maxwell Maltz*, Creative Living for Today, Gallery Books, August 1978

[62] Mandel (2008): *Ernest Mandel*, Einführung in den Marxismus, ISP, Köln, 2008

[63] Mar et al. (2006): *Raymond A. Mar, Keith Oatley, Jacob Hirsh, Jennifer dela Paz, Jordan B. Peterson*, Bookworms versus nerds: Exposure to fiction versus non-fiction, divergent associations with social ability, and the simulation of fictional social worlds, Journal of Research in Personality, Volume 40, Issue 5, October 2006, Pages 694-712

[64] Marvel Fandom - Victor von Doom (Earth-616): https://marvel.fandom.com/wiki/Victor_von_Doom_(Earth-616) *Accessed on 5. May 2023*

 Beyond Simplicity

[65] Marx (1859): *Karl Marx*, A Contribution to the Critique of Political Economy

[66] Marx (1990): *Karl Marx*, Capital Volume 1, Penguin Classics, 1990

[67] Martín (2023): *Ruth Martín*, Challenge based learning (CBP): The evolution of PBL? https://blog.genial.ly/en/challenge-based-learning/ *Accessed on 15. May 2023*

[68] Mattson (2015): *Mark P. Mattson*, Lifelong brain health is a lifelong challenge: From evolutionary principles to empirical evidence, Ageing Research Reviews, Volume 20, March 2015, Pages 37-45

[69] May et al. (2022): *Brienne K. May, Jillian L. Wendt, Michelle J. Barthlow*, A comparison of students' interest in STEM across science standard types, In. Social Sciences & Humanities Open, Volume 6, Issue 1, 2022

[70] More (1934): *Louis Trenchard More*, Isaac Newton — A Biography, Charles Scribner's Sons, 1934

[71] Miragliotta et al. (2018): *Giovanni Miragliotta, Andrea Sianesi, Elisa Convertini, Rossella Distante*, Data driven management in Industry 4.0: a method to measure Data Productivity, IFAC-PapersOnLine, Volume 51, Issue 11, 2018, Pages 19-24

[72] Murphy (1988): *Gregory L. Murphy*, Comprehending complex concepts, Cognitive Science, Volume 12, Issue 4, October–December 1988, Pages 529-562

[73] Newton (2018): *Derek Newton*, What Mark Twain Didn't Really Tell Us About Technology Disruption, Jobs And Education, Forbes, 2018 https://www.forbes.com/sites/dereknewton/2018/07/26/what-mark-twain-didnt-really-tell-us-about-technology-disruption-jobs-and-education/ Accessed on 08.05.2023

[74] NSTA (2014): *National Science Teaching Association*, Systems and System Models, https://ngss.nsta.org/CrosscuttingConcepts.aspx Accessed 15. May 2023

[75] OECD (2006): *Organisation for Economic Co-operation and Development Global Science Forum*, Evolution of Student Interest in Science and Technology Studies, Policy Report.

[76] Poulantzas (2014): *Nikos Poulantzas*, State, Power, Socialism, Verso, 2014

[77] Pop Culture Detective (2023): Disney's Strange Solarpunk World, *Pop Culture Detective*, https://www.youtube.com/watch?v=rqQJHja9qxU *Accessed on 9. May 2023*

[78] Purcell & Rainie (2014): *Kristen Purcell, Lee Rainie,* Americans Feel Better Informed Thanks to the Internet, Pew Research Center, December 8, 2014

[79] Rampersad (2020): *Giselle Rampersad,* Robot will take your job: Innovation for an era of artificial intelligence, Journal of Business Research, Volume 116, August 2020, Pages 68-74

[80] Reche & Perfectti (2020): *Isabel Reche, Francisco Perfectti,* Promoting Individual and Collective Creativity in Science Students, Trends in Ecology & Evolution, Volume 35, Issue 9, September 2020, Pages 745-748

[81] Renatovna & Renatovna (2021): Akramova Gulbahor Renatovna, Akramova Surayo Renatovna, Pedagogical and psychological conditions of preparing students for social relations on the basis of the development of critical thinking, Psychology and Education 2021, 58: 4889-4902

[82] Rietzler & Grolimund (2023): Stefanie Rietzler, Fabian Grolimund, Erfolgreich lernen mit ADHS: Der praktische Ratgeber für Eltern, Hogrefe AG, 2023

[83] Roepke et al. (2017): *Ann Marie Roepke, Lizbeth Benson, Eli Tsukayama & David Bryce Yaden,* Prospective writing: Randomized controlled trial of an intervention for facilitating growth after adversity, The Journal of Positive Psychology, Volume 13, 2018 - Issue 6

[84] Rubin (2015): *Gretchen Rubin,* Better Than Before: Mastering the Habits of Our Everyday Lives, Crown, 2015

[85] Second Thought (2023): What If We Just...Stopped Working? https://www.youtube.com/watch?v=2i0RrGx_GrE *Accessed 23. May 2023*

[86] Singh et al. (2019): *Rajni Singh, Devika, Christoph Herrmann, Sebastian Thiede, Kuldip Singh Sangwan,* Research-based Learning for Skill Development of Engineering Graduates: An empirical study, Procedia Manufacturing, Volume 31, 2019, Pages 323-329

[87] Smil (2022): *Vaclav Smil,* How the World Really Works, Penguin Random House UK, 2022

[88] Smith (1776) *Adam Smith,* Wealth of Nations, London, 1776

[89] Sládek et al. (2011): *Petr Sládek, Tomáš Miléř, Renáta Benárová,* How to increase students' interest in science and technology, Procedia - Social and Behavioral Sciences, Volume 12, 2011, Pages 168-174

[90] Sokal & Bricmont (1999): *Alan D. Sokal, Jean Bricmont*, Fashionable Nonsense: Postmodern Intellectuals' Abuse of Science, St Martin's Press, 1999

[91] Spraul (2012): *V. Anton Spraul*, Think Like A Programmer, No Starch Press, 2012

[92] Standard (2023): Desinteresse an Wissenschaft größer als Skepsis, https://www.derstandard.at/story/2000142268125/desinteresse-an-wissenschaft-groesser-als-skepsis, *Accessed on 2. May 2023*

[93] Starr (1919): *Mark Starr*, A Worker Looks At History, Plebs League, 1919

[94] Steen (1978): *Lynn Arthur Steen*, Mathematics Today: Twelve Informal Essays, Springer, 1978

[95] Süskind (1990): *Patrick Süskind*, Die Taube, Diogenes Verlag, 1990

[96] Tang et al. (2009): *Tang, H. E., Voon, L. L. & Julaihi*, N. H. A case study of 'high-failure rate' mathematics courses and its' contributing factors on UiTM, Sarawak diploma students. Paper presented at the Conference on Scientific & Social Research, 14-15 March 2009

[97] Taylor et al. (2004): *Lorraine C. Taylor, Jennifer D. Clayton, Stephanie J. Rowley*, Academic Socialization: Understanding Parental Influences on Children's School-Related Development in the Early Years, Review of General Psychology, Educational Publishing Foundation, 2004, Vol. 8, No. 3, pages 163–178

[98] Thompson (2022): *Karl Thompson*, Marxism and Culture, November 4, 2022 https://revisesociology.com/2022/11/04/marxism-and-culture/?utm_content=cmp-true Accessed on 13. May 2023

[99] University of York (2023): *Electronic Engineering, University of York*, Why study Electronic Engineering? https://www.youtube.com/watch?v=Xy4xf3SEwvM Accessed on 26. May 2023

[100] Vance (2015): *Ashlee Vance*, Elon Musk: Tesla, SpaceX, and the Quest for a Fantastic Future, HarperCollins, 2015

[101] Venkatesan et al. (2023): Ramnarayan Venkatesan, The Swamp Thing - The Parliament of Gears, DC Comics, 2023

[102] Wade (1834): *John Wade*, History of the Middle and Working Class, Second Edition, published by Effingham Wilson, 1834

[103] Wang et al. (2021): *Bojie Wang, Qin Zhang, Fengqi Cui*, Scientific research on ecosystem services and human well-being: A bibliometric analysis, Ecological Indicators, Volume 125, June 2021, 107449

[104] Weng et al. (2022): *Xiaojing Weng, Thomas K.F. Chiu, Cheung Chun Tsang*, Promoting student creativity and entrepreneurship

through real-world problem-based maker education, Thinking Skills and Creativity, Volume 45, September 2022

[105]   Wolters (2013): *Eugene Wolters,* Read Derridas Response to the Sokal Affair, 27. August 2013 http://www.critical-theory.com/read-derridas-response-sokal-affair/ Accessed on 14. May 2023

[106]   World Economic Forum (2016): The Future of Jobs – Employment, Skills and Workforce Strategy for the Fourth Industrial Revolution – Executive Summary, January 2016

[107]   World Economic Forum (2023): Future of Jobs Report – Insight Report, May 2023

[108]   Yang et al. (2018): *Zhi Yang, Ying Zhou, Joanne W.Y. Chung, Qiubi Tang, Lian Jiang, Thomas K.S. Wong,* Challenge Based Learning nurtures creative thinking: An evaluative study, Nurse Education Today, Volume 71, December 2018, Pages 40-47

[109]   Yun (2023): *Tao Yun,* Review of science and technology innovation policies in major innovative-oriented countries in response to the COVID-19 pandemic, Biosafety and Health, Volume 5, Issue 1, February 2023, Pages 8-13

[110]   Zeidler (2006): *Eberhard Zeidler,* Quantum Field Theory I: Basics In Mathematics and Physics, Springer, 2004

[111]   Zer0Books (2023): https://www.johnhuntpublishing.com/zer0-books/ *Accessed on 11. May 2023*

[112]   Žižek (2020): *Slavoj Žižek,* The Indivisible Remainder, Verso, 2020

Made in the USA
Monee, IL
07 July 2026